Studies in Logic
Volume 7

Fallacies
Selected Papers 1972–1982

Volume 1
Proof Theoretical Coherence
Kosta Dosen and Zoran Petric

Volume 2
Model Based Reasoning in Science and Engineering
Lorenzo Magnani, editor

Volume 3
Foundations of the Formal Sciences IV: The History of the Concept of the Formal Sciences
Benedikt Löwe, Volker Peckhaus and Thoralf Räsch, editors

Volume 4
Algebra, Logic, Set Theory. Festschrift für Ulrich Felgner zum 65. Geburtstag
Benedikt Löwe, editor

Volume 5
Incompleteness in the Land of Sets
Melvin Fitting

Volume 6
How to Sell a Contradiction: The Logic and Metaphysics of Inconsistency
Francesco Berto

Volume 7
Fallacies — Selected Papers 1972-1982
John Woods and Douglas Walton, with a Foreword by Dale Jacquette

Studies in Logic Series Editor
Dov Gabbay   dov.gabbay@kcl.ac.uk

Fallacies
Selected Papers 1972–1982

John Woods and Douglas Walton
with a Foreword by Dale Jacquette

© Individual author and King's College 2007. All rights reserved.

ISBN 978-1-904987-16-1
College Publications
Scientific Director: Dov Gabbay
Managing Director: Jane Spurr
Department of Computer Science
Strand, London WC2R 2LS, UK

Original cover design by Richard Fraser
Cover produced by orchid creative www.orchidcreative.co.uk
Printed by Lightning Source, Milton Keynes, UK

All rights reserved. No part of this publication may be reproduced, stored in a retrieval system or transmitted, in any form, or by any means, electronic, mechanical, photocopying, recording or otherwise, without prior permission, in writing, from the publisher.

To the memory of C. L. Hamblin

Reasoning Awry:
An Introduction to Woods and Walton, *Fallacies: Selected Papers 1972–1982*

DALE JACQUETTE

'Sweet smoke of rhetoric!'
William Shakespeare, *Love's Labours Lost*, Act III, Scene I, Line 59

'If it weren't for erroneous conclusions, these people would never arrive at any at all.'
Richard Russo, *Straight Man* (1997)

A Modern Classic Redux

John Woods' and Douglas Walton's superb studies of the informal fallacies are finally back in print. There are more recent, but to my mind no better, in-depth diagnoses of main types of fallacious reasoning than the volume here reissued for a new generation of readers. Woods and Walton's collection of papers on faulty inference provides an invaluable resource not only for the detection and criticism of selected fallacies, but more importantly for the general field of informal logic and critical thinking.

Elements for a Theory of Fallacies

Woods and Walton disavow any title to presenting a *theory* of fallacies in the full and proper sense of the word in the book. This is not just modesty, let alone false modesty, but a sound recognition that there is much solid work to do in meeting these requirements, and Woods and Walton quite reasonably have not set themselves so lofty a goal.

Nor would it be reasonable to expect a fully comprehensive approach to the informal fallacies to be developed over a period of ten years' writing on so many different fallacies and published in such a wide array of different logic and philosophy journals. What is remarkable is rather the degree to which the individual papers, written over a decade's span and brought together in the present volume, nevertheless represent a cohesive grasp of the kinds of trickery to which reasoning is susceptible. A just appraisal of what Woods and Walton achieve in this compilation of papers on the informal fallacies must include the extent to which they make significant contributions toward a theory of fallacies. What emerges unmistakably from the cumulative effect of their individual investigations of particular fallacies is the basis for a theory that must minimally come to terms with the salient features of misguided reasoning they have identified.

The essential ingredients for a theory of fallacies include a general analysis of what it means for an inference to be fallacious, a principle for identifying and systematically organizing whatever inferences satisfy the requirements, including all those which have been well-known and widely discussed from classical times, and a lavish discussion of particular fallacies with memorable illustrations and appropriate explanations of how reasoning goes awry in such instances and how the fallacies in each subcategory are properly identified, interpreted, and avoided. Woods and Walton's album of papers on the fallacies provides the foundations for such a theory in timeless exemplary fashion, even when the authors do not profess to fulfill one and all of these principal requirements. What we find in the volume accordingly is an outstanding field guide to the major flora and fauna of faulty inference, with an implied sense of completeness and order that has yet to be matched by later writers on the subject. The result of the ten years' collaboration reflected in the papers assembled here is all the more noteworthy as a result for its piecemeal pursuit of a large-scale program in fallacy research, the broad outlines of which are visible in Woods and Walton, a promise for a future theory of fallacies that still awaits complete fulfillment by informal logicians.

What I appreciate and admire about the Woods-Walton fallacies collection is precisely its lack of an over-arching theory. Rather than prematurely fitting their discussions of particular fallacies to the requirements of a favorite theory, stretching, bending and lopping off parts to fit a Procrustean bed that analyzes the concept by means of distorting oversimplifications, Woods and Walton treat every fallacy in its own terms and on its own merits. They use minimal formal symbolic apparatus when appropriate, and otherwise confine their examination to discursive analysis, dialectical context, linguistic and psychological impact, doxastic implications, probabilistic inference, and whatever else is clearly required. In the process, the function of good reasoning in contrast with fallacies is established, emerging in stages from page to page, as the authors explain how to arrive at justifiable conclusions, fixing and shifting burden of proof in contexts of ongoing argument construed as a dialectical exchange, in its relation to larger informal epistemic obligations and purposes, including well-reasoned judgments of truth and other values.

Strange Allure of Fallacious Reasoning

Why, however, are some logicians, philosophers, and rhetoricians preoccupied by the study of informal fallacies in the first place? The answer is manifold, of course — they do so for an assortment of illuminating reasons. Here are some of main motivations; fallacies fascinate because they are:

- Intrinsically interesting.

- Crucial for understanding correct reasoning by instructive contrast with its opposite.

- Useful for evaluating arguments in all disciplines, including logic and philosophy.

- Unlimitedly valuable in practical applications distinguishing good reasoning and guiding decision-making.

An *inference* or *argument* is standardly defined as a sequence of propositions (categorical declarative sentences), some of which are distinguished as premises or assumptions from the others thereby differentiated as conclusions by an *inference indicator*. Inference indicators in turn are any of a wide variety of terms that mark the effort to draw a conclusion from other propositions, notably 'thus', 'therefore', 'hence', 'it follows that', and the like.

This broad definition leaves it open that there can be both good and bad inferences. Indeed, there can be wretchedly awful arguments of indefinitely many descriptions, most of which do not belong to any distinct category and are so patently bad that no one is likely to fall into or be taken in by them. Some of these atrocious arguments are nevertheless more devious, difficult to detect, and, for a number of reasons, easy to confuse with correct forms of reasoning, on the part both of those who make them purposely or inadvertently, and those who succumb to their spell and allow themselves to be persuaded by reasoning that on more careful reflection they should and hopefully would reject as erroneous.

A *fallacy* is, again, standardly, defined as any logically incorrect argument. It is sometimes said that a fallacy is an incorrect argument that looks or can look to someone to be correct, or that appears to be correct, or that can easily be mistaken as correct. Human frailty and gullibility being what it is, however, this characterization is potentially true of absolutely any argument, correct or incorrect. The analysis of fallacies is most competently undertaken as a result by analyzing the kinds of errors identifiable categories of faulty reasoning are typically said to embody.

There are three ways in which an inference can embody a fallacy, based on the standard three-part definition of a correct argument. An argument, in order to be circumspect, must be (1) *deductively valid* or *inductively correct*; (2) *sound*, consisting of only true assumptions; and (3) *significant*, meaning, among other things, that is relevant, noncircular, and the like. An argument commits a fallacy whenever it fails to meet all three conditions. If we consider the validity or structural probity of deductive and inductive inferences separately according to their distinct definitions, then we can consider a simple taxonomy of three major categories of fallacies: (i) *fallacies of* (formal or informal) *deductive invalidity* or *improper inductive projection* (also known as *non sequiturs*); (ii) *fallacies of unsoundness* (concerning logical consistency and the logic of inference from evidence); (iii) *fallacies of insignificance* (involving logical irrelevance, circularity, *inter alia*).

Such a division of fallacies by type is by no means accepted by all logicians. There is in fact no universal method of classifying fallacies. Logicians disagree substantially about which types of inference are fallacious, and about how to arrange recognized fallacies into categories. There are advantages and disadvantages to any of the well-received proposals for organizing fallacies currently on the market. Symbolic and informal logic and critical reasoning theorists even disagree about whether all of the traditionally recognized types really are genuine fallacies, and, for that matter, about what exactly is or should be meant by the concept of a fallacy. One of the advantages of Woods and Walton's consideration of fallacies that makes this compilation of essays so enduring is that they do not limit themselves to any particular scheme of fallacies while saying just enough to suggest a variety of ways in which the fallacies they choose to discuss might be arranged. It is part and parcel of the study of fallacies as a chapter of informal logic that the fallacies, beyond being classified as violations of one or more of the three standard requirements of correct inference, do not readily lend themselves to any particular rigid categorization.

From ancient times, beginning with Aristotle's *De Sophisticus Elenchis, On Sophistical Refutations*, through the medieval manuals of logic and rhetoric, to such present-day digests as Woods and Walton, the fallacies have inevitably been deposited into a grab-bag of one sort or another. Often, there has been an attempt to impose some kind of order on the fallacies, but these efforts usually lack conviction and are undertaken more for the sake of convenience and ease of memory than as a natural taxonomy. Mnemonic devices and implausible taxonomies aside, one senses the difficulty of housing all of the informal fallacies together under one roof simply in considering the enormous differences between, say, the *petitio principii*, slippery slope, and gambler's fallacy; or between any of these and the *post hoc, propter hoc*, abusive *ad hominem*, *ad baculum*, *ad populum*, or *ad ignorantium* — once again, to pick at random from the bulging grab-bag. An underlying account that connects the dots between these apparently very different and distinct types of traditional fallacies there may someday be; but until such time the better course is surely to celebrate their diversity and try to cope individually with each unique genus and species as its peculiarities seem to require.

Fallacies Within the Field of Informal Logic

That informal as opposed to symbolic logic is the natural harbor of the appropriately designated informal fallacies is manifest in these and other facts about the fallacies themselves. Woods and Walton clearly articulate their attitude toward the differences between formal and informal logic. They explain why they consider their project as belonging squarely to what in 1989 was then the neophyte discipline of informal logic. Nor has the situation changed much today, although there are now departments, divisions within departments, and journals devoted to informal logic, and a recognizable informal logic movement within the more encompassing study of logic. Informal logic continues to struggle as a distinct subject seeking its identity alongside mathematical or formal symbolic logic.

In addition to the well-taken points Woods and Walton raise in their 'Introduction' in support of the fallacies belonging to the field of informal logic, there are considerations to show that symbolic logic is inherently inadequate to explicate the motivations for fallacies and their exact reasons for failing to satisfy the formal requirements of deductive validity. Thus, in the *ad baculum* fallacy, for example, the deductive error of the fallacy is obvious. From the fact that violence, or, less prosaically, some sort of ill effect, is likely to follow if the argument's conclusion is not accepted, it by no means follows logically that the conclusion is true. Fair enough, perhaps, as far as it goes. What is disappointing about the formal analysis of any such informal fallacy is, first of all, that it is formally, syntactically indistinguishable from the isomorphic formal analysis of virtually any other so-called informal or rhetorical fallacy in the grab-bag. Appeal to authority and *ad ignorantiam* or appeal to ignorance, poisoning the well, *ad populum*, and many (though notably not all) others of the rightly so-called informal or rhetorical fallacies. All have the deductively disreputable logical structure, when uninformatively reconstructed as: P therefore Q, or minor variants thereof. You will suffer loss if you accept the proposition that S; therefore

S is not true. Many people believe the proposition that S; therefore S is true.

What distinguishes the patently very different fallacies in this large family of fallacies is, consequently, invisible to purely formal deductive logic. Once symbolic logic has satisfied itself that a given fallacy in this group is deductively invalid by virtue of its misbegotten inferential *form*, formal symbolic logic has nothing further of interest to say about the *content* of such fallacies. Formal logic alone is altogether silent about why it is that thinkers are sometimes inclined to commit fallacies of any chosen type, and why those on the receiving end of arguments are sometimes persuaded by what is undeniably a deductively invalid inference. The main deficit of purely formal analysis of the fallacies that as a result have deserved their traditional classifications, is that they do not distinguish any of the fallacies by type. However, each of the classical fallacy types has its own seductive appeal, which a complete analysis of why it is fallacious must somehow adequately address. Formal logic is blind to the differences that distinguish any category of informal fallacy from any other, or from such simply dull-witted deductively invalid inferences as 'Snow is white, therefore, God exists'; 'Snow is white, therefore, grass is green', and unlimitedly many others, all of which equally fit the formal pattern of the deductive *non sequitur*.

The problem is even more eloquently and definitively illustrated by the widely practiced and frequently difficult to detect fallacy of *equivocation*. The fallacy is superficially perpetrated when someone reasons in what appears formally to be a perfectly valid inference, such as *modus ponendo ponens*, while trading on at least two different meanings attributed to at least wo different token occurrences of a syntactically identical term, or, colloquially, *homograph*. Suppose I think or say: If I am carrying a bow, then I can shoot an arrow; I am carrying a bow; therefore, I can shoot an arrow. The argument gets (partially) formalized, typically, as If P, then Q; P; therefore, Q. This is perfectly deductively valid. Suppose, further, now, that by the key term 'bow' in the first assumption, I intend a weapon, and in the second assumption by the homographic term 'bow' I intend a piece of ribbon shaped and tied in a certain way or a device made with horsehair for playing a stringed instrument. The inference in that case is evidently deductively invalid; yet the tools needed to uncover the mistake, to expose the fallacy of equivocation it contains, are evidently not a matter of logical syntactical form to which formal symbolic logic is limited.

Another aspect of informal fallacies far removed from formal logic's myopic view of faulty inference is the fact that most of the fallacies have an intuitive appeal by which they catch the unwary. Consider the *tu quoque*.

It seems reasonable, at least to a certain extent, to respond to persons who object on principle to my failure to attend a meeting that they were not there either, and that they have frequently been absent from similar meetings in the past. Why should I accept the advice and moral admonitions of persons who cannot even follow it themselves? Why should I agree with the conclusions of a sermonette about the virtues of voting in a general election delivered by someone who has never even registered to vote? Of course, the sermonizer might be right, despite not practicing what he or she preaches, and the argument in support of doing one's civic duty might be correct regardless of its source and the accidents of the source's biography. Still, when I reply *tu quoque* to the preacher, pointing out that he or she has not voted, am I not arguing correctly? Would I not be right to turn the argument around on its author, and would the author not be right to take the lesson to heart and decide to vote?

It can even come to seem reasonable to use force in the *ad baculum* as a means of persuasion when the cause, say a life-or-death struggle against fascism, is just, and no other means of persuasion is available or likely to succeed. As to the *ad verecundiam* or appeal to authority, which Woods and Walton discuss in several of their papers, it is surely reasonable to proceed in many if not most contexts of argument on the basis of what we believe experts to have concluded is true. Otherwise we must all become professional physicists and genetic scientists and lawyers and meteorologists, and so on, indefinitely, in order to reason properly about any of the things these experts have to say; and even then we will be relying on our own testimony as experts in all such argument contexts. That we must judge the expertise of purported experts judiciously goes without saying, and that the evidence taken under consideration in drawing inferences from experts can at best lend a high probability to the truth of what they say and the conclusions we infer must also be acknowledged. There is nonetheless a vast difference between the sense in which an appeal to (genuine) authority is fallacious and, say, denying the antecedent or affirming the consequent of a conditional is fallacious, or even in comparison with such traditional fallacies as the sorites, heap or bald man, many questions, or *post hoc, propter hoc* fallacy. Fallacies command attention in some instances because they appeal to factors that we must sometimes rely on in legitimate argument, but that can lead us astray if we do not exercise appropriate caution and avoid excess or blind reliance on all inferences of their sometimes valid, sometimes invalid forms. If, indeed, it is true that form alone in such arguments by itself is not enough to distinguish an inference as valid or fallacious, where this is precisely the feature that seduces the reckless reasoner, it is clear that formal logic will be powerless to sort out the wheat from the chaff. The same holds to an even greater degree where probabilistic and other types of inductive reasoning are concerned, as seems to be the case for applications of the *ad verecundiam*, whether warranted or fallacious.

It is not that fallacies charm simply by throwing dust in our eyes. They do not overmaster reasoning by exploiting a foolish desire to be manipulated, like Phyllis riding Aristotle piggyback in a medieval tapestry. Rather, certain types of fallacies deceive us by virtue of the fact that at some level and in some way they appeal to our rationality despite the fact they are strictly deductively invalid or inductively incorrect. There is something that could be right, that is partly right or seems to be right about the most persuasive fallacies. It is their strength and the source of greatest danger. They do not just superficially resemble good arguments; they possess to a limited degree the features that make good arguments reasonable. And yet they generally fail as good arguments in ways that are imperceptible to the methods of formal symbolic logic.

Nor are such formally mistaken types of fallacy as *petitio principii* adequately diagnosed by purely formal symbolic logic. These all ultimately have a form equivalent to P, therefore, P. The challenges encountered by formal logic in understanding this family of informal fallacies are equally apparent. Not only are there question-begging arguments that are not readily translated into anything quite so blatantly circular as this, but, more importantly, we have the objection already in J.S. Mill's *A System of Logic* that every deductively valid argument embodies circular reasoning. The famous example is the elementary Aristotelian syllogism in BARBARA: All men are mortal; Socrates is a man; therefore, Socrates is mortal. If we do not already know that Socrates is mortal, as the conclusion states, we could not possibly be in a position to support the truth of the assumption that *all* men are mortal. The objection in effect is that in the case of any deductively valid inference we must always the assume the truth of the conclusion in order to be able to uphold the truth of at least one of the inference's assumptions. Exactly what is or should be meant by 'informal' logic, if it is not simply take by virtue of its grammatically negative terminology to be logic that is not formal, remains a difficult question for this general area of rhetorical-philosophical research. What one learns in part, among other things, of course, from Woods and Walton's selection of papers on fallacies, is that the fortunes of informal logic are inextricably intertwined with systematic progress toward an intuitively correct understanding of the informal fallacies.

Navigating the Realms of Fallacy

The authors cultivate a salutary awareness of what it is the study of fallacies should and should not try to accomplish. They have a sharp sense of the limits beyond which inquiry into the fallacies strays at its own peril, of the methodology to be developed and employed in discussing each distinct type of fallacy, and the scope and limits of the formal and informal modes of analysis deployed in understanding every category of fallacious reasoning in their catalog.

Woods and Walton, in keeping, most probably, with an unacknowledged concurrence with the grab-bag model of the informal fallacies, do not propose to discuss every fallacy that has ever made an appearance in discussions of rhetorical excess. What is presented here are their papers on fallacies published during the ten years time covered by the volume's subtitle. They do not include some interesting fallacies that it would be interesting to have their opinions about, however. These comprise, among others, the slippery slope; sorites; intensional fallacy; naturalistic fallacy, false alternatives; use-mention confusion; hasty generalization; *ad consequentiam; ad misericordiam; ad odium; ad naturum; ad superbium; tu quoque;* guilt by association; poisoning the well; genetic fallacy.

Woods and Walton also judiciously exclude a myriad of specialized fallacies that seem to crop up whenever a philosopher encounters a style of reasoning that seems inadequately justified or at odds with a preferred position. Such 'fallacies' are arguably too freely named as such and too particular to a philosophical methodology or ideology, not to mention, for that very reason, too controversial *as* fallacies in the ordinary sense, to warrant inclusion in a general exposition on the illogic of common types of fallacious reasoning. Woods and Walton appropriately refrain from including these from their expositions. We may nevertheless apply the lessons of their analyses to what purport to be fallacies in these further categories. It is a valuable exercise, upon completing the anthology, to try filling in the blanks left by Woods and Walton in Woods-and-Walton-esque style, making an educated guess as to what the authors might have had to say about these additional fallacies, given what their approach to the fallacies they choose to discuss suggests.

The papers by Woods and Walton presented here have stood the test of time. They were originally published in journals or in one case as a chapter in the proceedings of a conference on informal logic, they were anthologized in the first publication of the book, and now they are reprinted and reissued in this new edition. The republication of this invaluable volume must be a welcome event and cause for celebration by all theorists and practitioners of good reasoning.

Final Note

The essays by Woods and Walton reprinted in this collection appear in *chronological* order of their first publication between the years 1972 and 1982. This is a sensible plan; but readers who prefer a more thematic itinerary might consider the following sequence, which I find useful in approaching the subject matter from a topical perspective.

One might begin with the Introduction, then turn to Chapter 17 'What is Informal Logic', proceeding to Chapter 1 'On Fallacies', Chapter 15 'Equivocation and Practical Logic', and proceed to a series of chapters on circular reasoning, begging the question or *petitio principii*, beginning with Chapter 3 '*Petitio Principii*', Chapter 6 '*Petitio* and Relevant Many-Premissed Arguments', Chapter 10 'Arresting Circles in Formal Dialogues', Chapter 19 'Question-Begging and Cumulativeness in Dialectical Games', Chapter 12 'Circular Demonstration and von Wright-Geach Entailment', and Chapter 13 'Laws of Thought and Epistemic Proofs'. From this point, it may then be useful to consider papers on specific classical paradoxes. Here one might choose Chapter 2 '*Argumentum ad Verecundiam*', Chapter 14 'What Type of Argument is an *Ad Verecundiam*?', and then advance to Chapter 4 '*Ad Baculum*', Chapter 5 '*Ad Hominem*', Chapter 7 '*Ad Hominem*, Contra Gerber', Chapter 11 'The Fallacy of *Ad Ignorantiam*', Chapter 16 'Why is the *Ad Populum* a Fallacy?', Chapter 9 '*Post Hoc, Ergo Propter Hoc*', and concluding with Chapter 8 'Composition and Division', and Chapter 18 'The Fallacy of Many Questions'. More concisely and schematically, the recommendation is to consider also reading the essays in this order:

$$\text{Intro} \to 17 \to 1 \to 15 \to 3 \to 6 \to 10 \to 19 \to 12 \to 13 \to 2 \to 14 \to 4 \to 5 \to 7 \to 11 \to 16 \to 9 \to 8 \to 18.$$

Other possibilities will also occur to readers, pursuing special interests in particular topics, or wishing to build on background reading in related fields before tackling essays dedicated to more technical aspects of evaluating informal fallacies. The many choices available for a reading program threading through the essays in different ways is only one of the riches and delights of Woods and Walton's rewarding work on *Fallacies*.

Contents

Preface	xiii
Introduction	xv

Chapter 1. **On Fallacies** — 1

1. Fallacious Arguments — 1
2. The Concept of Argument — 2
3. Deductive Arguments — 3
4. Beliefs — 4
5. Epistemic Logic — 5
6. Probabilistic Models — 6
7. Dialectical Precepts — 8

Chapter 2. *Argumentum ad Verecundiam* — 11

1. Arguments from Expertise — 11
2. Stochastic Explanation — 14
3. The Logical Structure of the *Ad Verecundiam* — 15
4. Adequacy Conditions — 17
5. The Analysis Elaborated — 21
6. Dialectic — 25
7. Privileged Access and Expertise — 26
8. Infallibility — 27

Chapter 3. *Petitio Principii* — 29

1. Equivalence and Dependency Conditions — 29
2. The Disjunctive Syllogism — 32
3. Immediate Inference — 35
4. Universal Arguments — 39
5. Circular Definition — 42
6. Effectiveness and Soundness — 43

Chapter 4. *Ad Baculum* — 47

1. The Statement Requirement — 47
2. Prudential Inference — 48
3. *Ad Baculum* as Equivocation — 50
4. Threats and Beliefs — 52
5. Conclusions — 53

Chapter 5. *Ad Hominem* — 55

1. The Lockean View — 55
2. The Standard View — 57
3. The Circumstantial *Ad Hominem* — 58
4. Logical Inconsistency — 59
5. Assertional Inconsistency — 62
6. Praxiological Inconsistency — 63
7. Deontic-Praxiological Inconsistency — 63
8. The Abusive *Ad Hominem* — 65
9. Expertise — 65
10. Nexpertise — 66
11. Cranks and Crackpots — 70
12. Some Further Interconnections — 71
13. Concluding Remarks — 72

Chapter 6. *Petitio* and Relevant Many-Premissed Arguments — 75

1. *Petitio Principii* — 75
2. Mill: The Syllogism as *Petitio* — 77
3. De Morgan's Rejoinder — 78
4. Relevant Arguments — 80
5. Disjunctive Syllogism — 81
6. Conjunctive Conclusions — 83
7. Concluding Remarks — 85

Chapter 7. *Ad Hominem*, Contra Gerber — 87

1. Desiderata — 87
2. Neutral and Abusive Terms — 88
3. Laudatory *Ad Hominem* — 89
4. Lockean *Ad Hominem* — 90
5. Concluding Remarks — 91

Chapter 8. Composition and Division 93

1. The Mediaeval View 93
2. The Modern View 97
3. The Two Views Juxtaposed 99
4. The Basis of the Fallacy: Aggregates 101
5. Analysis of the Fallacy 110
6. The Cause of the Fallacy 113
7. Importance of these Fallacies 115
8. Concluding Remarks 117
   Appendix A 119
   Appendix B 119

Chapter 9. *Post Hoc, Ergo Propter Hoc* 121

1. The Problem 121
2. Basic Characteristics of Causation 121
3. Structure of *Post Hoc* Arguments 129
4. Seven *Post Hoc* Fallacies 132
5. Completeness 138
6. Correct and Incorrect Causal Arguments 139

Chapter 10. Arresting Circles in Formal Dialogues 143

1. Background 143
2. The Game H 144
3. Circle Games 147
4. Adequacy of H+(W)+(RI) for Representing the *Petitio* 149
5. Cumulativeness 153
6. Groundedness 155
7. A Linear Simplification 157
8. Concluding Remarks 158

Chapter 11. The Fallacy of *Ad Ignorantiam* 161

1. Find the Fallacy 162
2. Confirmation: Dis and Un 163
3. A Fallacy of Confirmation 164
4. An Epistemic Fallacy 165
5. A Dialectical Fallacy 166
6. The Analysis 167
7. Presumptions 168

    8. Testability   169
    9. Addendum   170

Chapter 12. Circular Demonstration and Von Wright-Geach Entailment   175

    1. Circularity Conditions   175
    2. The Von Wright-Geach Definition   176
    3. Revising the Definition   177
    4. Refinements   178

Chapter 13. Laws of Thought and Epistemic Proofs   181

    1. Moorean Proof   182
    2. Epistemic Assumptions   183
    3. Non-Classical Connectives   184
    4. Indirect Proof   187
    5. Circular Proofs   188
    6. Conclusions   190

Chapter 14. What Type of Argument is an *Ad Verecundiam*?   191

    1. Type of Argument   191
    2. Two Characteristics   192
    3. Neither Inductive Nor Deductive   192
    4. Plausible Argument   193

Chapter 15. Equivocation and Practical Logic   195

    1. The Nature of the Problem   196
    2. Contextual Shift   198
    3. Varieties of Equivocation   200
    4. Gross and Subtle Equivocations   203
    5. Hamblin: More Problems about Meaning   203
    6. Argumentation in Natural Languages   205
    7. Concluding Remarks   207

Contents   xi

Chapter 16.   **Why is the *Ad Populum* a Fallacy?**   209

    1. Arguments Directed to Specific Recipients   210
    2. *Ad Populum* as an Emotive Fallacy   213
    3. Is it an Argument at All?   214
    4. Is it a Single Fallacy?   217

Chapter 17.   **What is Informal Logic?**   221

    1. Applicability of Formal Methods   221
    2. Contextual Peculiarities   224
    3. Fallacies   227
    4. Illustration of Using Formal Theories: The Fallacy of Composition and Division   228

Chapter 18.   **The Fallacy of Many Questions**   233

    1. Semantic Preliminaries   234
    2. Three Major Methods   235
    3. A Pragmatic Approach   241
    4. Epistemic Settings   242
    5. Loaded and Multiple Questions   244
    6. Concluding Remarks   249

Chapter 19.   **Question-Begging and Cumulativeness in Dialectical Games**   253

    1. The Hamblin Game (H)   254
    2. Rescher Dialectics   259
    3. The Mackenzie Game DC   261
    4. Begging the Question in DD   267
    5. Organizing retractions of commitments   269

Notes   273

Bibliography   289

Acknowledgements   307

Index   311

Preface

We are gratified by the interest shown over the years in our writings on the fallacies. Readers or would-be readers have not always had an easy time in locating these writings, many of which are scattered in issues of journals not available in all libraries. Consequently it has frequently been suggested that our papers be collected into a single volume and, on the assumption that their half-life would exceed a thirty minute chat in the common room, we welcome the opportunity to republish afforded us by the series editors.

It has been decided to organize the volume as follows:

1. The papers will be set forth in a loosely chronological order.

2. The papers recur here with minimal adjustment and, so appear with their original imperfections, undisturbed by hindsight. We do not particularly relish displaying our early over-simplifications of complex problems, but think that doing so is warranted by two consideration. First, some of the over-simplifications are instructive; and, second, arranged in their pristine form, the papers will show, with some accuracy, a certain development in our views of the fallacies and our methods for dealing with them.

3. We have tried to excise prefatory and expository repetitions where we judge these to be an annoyance and distraction for the reader.

4. Owing to limited space, we have decided not to include our post-1982 papers. Should they continue to have a "shelf-life" over the next few years, it may be appropriate to consider their republication at a later date.

With the exception of three chapters, these papers were co-authored by Woods and Walton. Chapter 17 was written by Woods alone, and chapters 16 and 18 by Walton alone.

To the editors and publishers of the publications in which the material originally appeared, we express our thanks for permission to republish, generously given in every case. We also thank Brent Hudak, for his able assistance as Woods' research associate in 1986-87; Professor I.D.C. Newbould, Dean of Arts and Science, the University of Lethbridge, who supported the research appointment financially; and also the Research Fund of the University of Lethbridge and the Social Science and Humanities Research Council for financial assistance; Mrs. Candace Petek for excellent secretarial help; Ms. Julia Dennis, Wood's research associate in 1988-89, for her tenacious proofreading; the Institute for Functional Research of Language and Language Use (IFOTT) in Amsterdam for the use of their computer equipment; Dr. E.T. Feteris, Bart Garssen and Lisa Stevens, of the University of Amsterdam, for hazarding the book through the presses and for compiling the index; and Professor Frans van Eemeren and Dr. Rob Grootendorst, both of the University of Amsterdam and editors of this series, for generous and able support and advice. We would also like to acknowlege support of Walton's research by the Killam Foundation of the Canada Council through its award of a Killam Research Fellowship, and the support of the Netherlands Institute for Advanced Study in the Humanities and Social Sciences through its award of a Fellowship-in-Residence, and the Social Sciences and Humanities Research Council of Canada through the award of a Research Grant on Pragmatics of Argumentation.

John Woods, Lethbridge

Douglas Walton, Winnipeg

December, 1989

Introduction

In the early seventies, having arranged to meet at the Vancouver Airport, we completed the journey to the American Philosophical Association's Pacific Division meetings in San Francisco. As we recall, we took a Western Airline flight. This company was heavily promoting its Vancouver-U.S. routes, and the promotional medium was champagne - all you wanted. By the time we reached the San Francisco airport, we had concluded that the informal fallacies, notwithstanding some excellent work here and there, were in deplorable shape, and we were relaxed enough to think that the two of us jointly might venture to ameliorate that situation, however modestly. So we began, at whose initiative we could not now say, a fifteen year collaboration which has brought us much satisfaction both personally and professionally, and from the first ten years of which we have drawn the contents of this volume.

It would be naive to think that people who work on informal logic have a single, simple rationale for doing so, or that they pursue common targets, employ settled and well-approved methods, or that they are moved by the same dissatisfactions with how logic is done elsewhere. Still, various metalogical themes are discernible in much of the recent writings on informal logic, some of which are worth noting. One can for example recognize the following convictions, held with varying degrees of intensity and confidence, not all of which are entirely well-explained or well-founded. But they are there.

One hears that formal logic is *too narrow* to model adequately anything as rich and complex as actual argumentation. For formal logic is still basically the theory of deduction, restrictively conceived in simple sentential, first order and modal environments.

Formal logic is *too abstract*, as witness first order model theory in infinite domains.

Formal logic squanders too much of its energies on merely technical questions - on the technical *tour de force* - such as the representation of truth-functionality *via* the Sheffer-stroke.

Formal logic is *too metatheoretical,* as witness its affection for such things are the Compactness Metatheorem.
Formal logic is *too counterintuitive*, sacrificing intuitiveness to algorithmicity, as witness its treatment of implication and equivalence.

Fuelled by disapproval, a good deal of informal logic is crochety, suspicious, antitheoretical, doctrinaire and apocalyptic, and thus in many ways we think, wrong-headed and precipitate. Nevertheless, there is something of genuine value to these demurs, and it derives, we think, from the overarching *pedagogical* belief that formal logic is too this-that-and-the-other thing for the purpose of an accurate and intellectually satisfying *introduction* to the subject. It is not at all surprising that so much current work in informal logic is to be found in those introductory texts and articles which very obviously reflect this same instructional conviction.

We admit to thinking that there has occurred some misunderstanding of our own approach to these matters. No doubt we have, owing to obscure exposition, invited certain misconceptions. They are, we think, important enough to warrant explaining them away. We illustrate with reference to how we go about our work in fallacy theory.

1. We do not, for example, hold that fallacy theory, which is where most of our joint efforts fall, is all or even the central part of informal logic or critical thinking. Those familiar with our joint, and also our independent work[1] on the logic of dialogues will know that though dialogical games are used as a medium for the treatment of some fallacies, this work is intended to be philosophically interesting in its own right, entirely apart from any application to fallacy theory.

2. Though we often reach for insights into the fallacies within theoretical structures, we do not think that there is yet anything like a mature *theory* of the fallacies or, for that matter, of any single fallacy. It seems to us that there is presently far too much unsettledness about preanalytic intuitions. The more urgent work for now is to clear away

the pretheoretical rubbish which obscures fundamental insights and impeded coherent theoretical effort.

3. We are not fanatical formalists. We do not think that formal methods and formal elaboration guarantee sounder philosophical accomplishment; nor do we think that informal logic informally approached is a constitutionally inferior intellectual product.

4. We are no enemy of applied logic. We applaud the utility of manuals of self-help for the ratiocinatively insecure. But we deplore application recipes that cannot imaginably work, such as the breezy and confident presentation of the propositional calculus as logical innoculation against the endemic and vicious lunacies which surfeit the pages of what pass for our daily newspapers. Equally, however, we realize that any better guide, if it is to give wise counsel about *how* to avoid thralldom to bad argument, must depend on knowing *what* is to be avoided and *why* it should be. And if this reserves application until there is some sound theoretical development, so be it.

In our writings on the fallacies, we have tended toward a methodological pragmatism. If we are methodologically versatile and flexible, we *also* try to steer clear of ideological preconceptions about the "nature" of the enterprise that we call informal logic. We resist, not always successfully (see chapter 17), disputations about whether informal logic, by its nature, admits of formal treatment or, for that matter, of theoretical manipulation of any sort. In our view, the "nature" of informal logic is not as yet much known, and there is therefore, little room for methodological apriorism. And so, we are prepared to wait and see. Comparatively untroubled by ideological distractions, we have tried to use formal methods or borrow from existing formal theories where doing so may produce results, even very tentative results such as the raising of new questions. It is a welcome by-product that, as tools of fallacy theory, intuitionistic logic, relatedness logic, plausibility theory, graph theory, aggregate theory and the like are richer vehicles than even their creators may have intended or noticed, and of deeper philosophical interest.

We are well aware that our work on the fallacies has not attained the settled and solid maturity of, say, Church's *Introduction to Mathematical Logic*. This is so partly because we are we, and Church is Church. But it also has something to do with the sprawling difficulty of the subject (much more complex and unyielding than a theoretically deep, mathematically convincing explication of the concept of deductive correctness, for example). It is also a result of the subject's frightful neglect for over a century, its lack of establishment respectability and its concomitant failure to attract the attention of larger numbers than it has yet done of our best philosophical minds. Informal logic has a great deal of growing up ahead of it. For example it has been generously said that there is nothing quite like *Arguer's Position*.[2] It is said to be a valuable piece of work which in various ways stands alone for now. If it sets new standards, it also raises more tough questions than it answers definitely and convincingly. It may be that it mistakenly takes the multi-person dialectical exchange to be primary, with easy adjustment for soliloquial cases, putting n=1. In fact, solo reasoning, as Robert Binkley calls it, may require a more elaborate treatment than this and may deserve its own paradigmatic place in theory. Similarly, the paper, "Composition and Division"[3] which some critics regard highly, is an imperfect and not fully mature effort. It is programmatic and tentative. It does not persist long enough with its central contention that aggregate theory gives the account of the part-whole distinction needed for the analysis of composition and division. It is better in its dismissive moments, and convincingly dispatches set theoretic, mereological and Noll-Suppesian models of the part-whole relationship. But it teases out only a few nomic themes of only shaky generality; for example, that not every predicate is compositionally hereditary.

As another example, the *ad verecundiam* is usefully approached via plausibility screening; but plausibility screening is a very limited device, leaving unexplained how, in a clash of expert opinions, initial assignments of expertise-rankings are justified.

Then, too, in "New Directions in the Logic of Dialogue", the suggestion is made that a "distinction between semantics as a truth-theoretic framework on the one hand, and pragmatics as a dialogue-theoretic framework on the other" can furnish a conciliatory answer to the question of how the formal erotetic frameworks of those such as Belnap, Aqvist and Hintikka can be applied to "help resolve the pragmatic problems of question-answer management in games of dialogue".[4] This is suggested, not proved. It is concretely elaborated only up to a point and never really to the point of lawlike pronouncement. "New Directions" is more promissory than a *fait accompli*. But, again, this feature of the work we think derives principally from the conceptual inchoateness of the subject matter.

Introduction   xix

In a series of works, many re-printed here, we have concentrated on the *petitio principii*. Some critics have kindly averred that the *petitio* is now better understood than it has ever been, and that it stand nearer to definitive closure than it did fifteen years ago. But it is not there yet; and is so partly because informal logic has not yet developed stable and comprehensive routines of proof and metaproof. It may be that the logic of the fallacies and dialogic are not amenable to proof-theoretic treatment in anything like the way that first order, modal and set theoretic systems are. But we believe it to be a prime deficiency of informal logic, and therefore of our own work, that it has not made enough headway in developing convincing methods of constructing and validating solid generalizations, even at the more practical and contextually sensitive levels appropriate to its subject matter.

There is a further difficulty. Informal logic offends establishment preference owing to its interdisciplinary character. To be at all at home with informal logic, it is well to know one's way around rhetoric, probability theory, statistics, prudential argumentation theory, doxastic logic, doxastic psychology, erotetic logic, assertion logic, game theory, semiotics, dialectic, decision theory, much of classical mathematical logic, and more besides. Some hint of this intellectual complexity and its concomitant burden of learnedness is given by *Argument: The Logic of the Fallacies*.[5] There one finds first order logic, dialectic, inductive logic, probability theory, relatedness logic, plausibility logic, game theory, aggregate theory, dialectic, and decision theory. Of this work it is has recently been said that "the ;best defense of fallacy theory we are aware of ... is the work of Woods and Walton ... in their textbook. *Argument: The Logic of the Fallacies*"[6]; and "we are inclined to say that no criticism of the fallacy theoretic approach to argument analysis can afford to fail to come to grips with this account of the logic of the fallacies".[7] It should be re- marked, however, that *Argument: The Logic of the Fallacies* is a textbook designed for first year courses in logic and critical thinking. It suffers, therefore, from the omissions, over- simplifications and superficialities of the beginner's text, to say nothing of the authors' expository and critical imperfections. Though it is very pleasant to hear it, saying that this book is currently at the top of fallacy theory is to damn the field of fallacy theory with faint praise. It shows, if the critical estimate is fair and sound, the state of underdevelopment of informal logic in 1987.

The papers, reprinted in this volume are first generation contributions to a revived interest in the fallacies. We are gratified to know that they are thought to have some initial and continuing value. But, as we have been saying, their further significance is their indication of how much there is still to do. We intend to do some of it, and we make so bold as to hope that ideological squabbles might for a while be postponed; that a tolerant and responsible methodological pluralism might be embraced; that the lingering distrust of interdisciplinary approaches might be abandoned; and that demanding and solid doctoral programmes in argumentation theory and fallacy theory might be fashioned and first rate people recruited into them.

For lack of space, we have decided not to include the *American Philosophical Quarterly* winning entry in the 1985 Prize Essay Competition, "Are Circular Arguments Necessarily Vicious?". As Nicholas Rescher has said, the topics for the Prize Essay Competition are chosen when they present issues that should be central and important in philosophy, but have not received the sustained effort of study they deserve. Not only the *petitio* but the whole area of the informal fallacies could be put into this category.

There are signs that this benign neglect of the subject is righting itself. Work in informal logic abounds, and important new journals are being established. One thinks immediately of *Informal Logic* at the University of Windsor and *Argumentation* at the Free University of Brussels and the University of Amsterdam.

Important centres of research are emerging. E.M. Barth and Erik Krabbe work in Groningen; Frans van Eemeren and Rob Grootendorst in Amsterdam; Kuno Lorenz in Saarbrücken; Michel Meyer in Brussels; Jaakko Hintikka in Tallahasee; Laurie Carlson in Helsinki; Ralph Johnson and Anthony Blair in Windsor; and David Hitchcock in Hamilton; Trudy Govier in Calgary and Jim MacKenzie in Sydney. These researchers, among many others, are producing work of increasing maturity and significance. There are also interesting efforts in faculties of education concerned with critical thinking, and in departments of speech communication concerned with the empirical study of argumentation.

Works of such scope and variety need to be transmitted across disciplinary boundaries and happily this has occurred at the two Windsor Conferences on Informal Logic and, more recently, at the Amsterdam Conference on Argumentation, and in many other places as well.

Introduction xxi

Yet another developing partnership is that between argumentation theory and artificial intelligence theory. AI examines structures of plausible reasoning and reasoned interactive dialogue between participants in question-answer argumentation. Hans van Ditmarsch, in Utrecht, is exploring the relationship of Expert Systems to argumentation theory. William C. Mann, of the University of Southern California, has noticed a growing interest in argumentation among several of the younger workers in AI research. AI techniques, in expert systems especially, bear many similarities to methods currently used to analyse argumentation in informal logic.

We close this Introduction on a semantic note. Readers will perhaps have noticed our acquiesence in the usage of "informal logic", not withstanding our artful and poignant discouragement of the practice (Chapter 17). However, the expression has taken root and does some responsible semantic work, and it would be churlish of us to persist in its repudiation.

Chapter 1
On Fallacies

Professor Hamblin's book is a welcome contribution to an abnormally underinvested domain of philosophy. Here is a detailed, scholarly, review of virtually the whole history of the theories of fallaciousness, and its abounds in pleasant truths and fascinating rediscoveries. Three chapters, as well, are given over to an original account of argument and dialectic of the author's own construction. It is bemusing that nearly everybody has had such profoundly dim things to say about fallacious argument, though this *does* tend to force one into a rather larger toleration of the inchoateness of the logic textbooks on the concept of argument and the informal fallacies. And it is understandable, therefore, that Hamblin's own account should exhibit, even grievously so at times, certain deficiencies and untowardnesses.

Our restricted purpose here is wholly elenchic. We mean to challenge Hamblin's account in the hope that exposure of its difficulties (which are more difficulties of the problems discussed than of their discussion) will contribute to the eventual emergence of a concept of argument more adequate to the domain of natural argumentation and of informal fallacies than the purely syntactic proof-theoretic accounts that, by themselves, are appropriate only to the domain of rigorous mathematical demonstration.

1. FALLACIOUS ARGUMENTS

"A fallacy is a fallacious argument". [80, p. 224] Well, aside from the fallaciousness of "Have you stopped beating your wife?" and such like, perhaps this may be so. Graver reservations are wanted for the suggestion that a fallacious argument must "seem valid but is not". [80, p. 224] The danger in such a proposal is that fallaciousness ultimately be construed as much a *psychological* attribute as a semantic one, with ⌜seems to v to be valid⌝ bearing the main freight. And with v a variable ranging over seemers and with seemers being what they are, one is beckoned right from the start by a thoroughgoing relativism in which virtually any given argument is both fallacious and not. Now, it is not that such a relativism is preposterous, but rather that it is a prodigally heady tonic for the ills to which Hamblin is sensitive. No one will deny that much of what is troublesome, even dangerous, about fallacies is that they so often go undetected. But this is true also of falling rocks; yet, who would wish to say that a rock is not falling unless it seemed to be at rest?

We might also note that by "valid" Hamblin does not mean "deductively valid". ⌜A, ∴ A⌝ is by his lights *not* valid. But then, since there is only one respect in which this argument even *seems* valid (viz. deductively), it is a mistake for Hamblin to suppose ⌜A, ∴ A⌝ fallacious; for though invalid, it does not seem otherwise. ("Validity", by the way, is undefined; it appears neither in the Index nor the Table of Contents.)

2. THE CONCEPT OF ARGUMENT

Given that a goodly part of the task at hand is to explicate the notion of a *fallacious* argument, it is entirely sensible to pursue a clear understanding of what an argument is. Hamblin is probably right in saying that a measure of the goodness of a theory of argument is the success it commands in dispatching Mill's *petitio* charge. [80, p. 229] Hamblin believes [80, p. 230-31] that a decent theory of fallacious argument requires the divorce of the notion of argument from what he calls "Formal Logic" (nowhere defined). He is struck [80, p. 226] by the habit of recognizing inductive arguments (valid and not), and of arguments *from authority* (bad and perhaps also good). He has doubts that this is a good habit. Whether or no inductive arguments and arguments from authority *are* arguments, one will clearly want to stop short of venturing either that a bad argument is not an argument or that a good argument is good in every relevant respect. All that is really needed for such restraint is one or other view of argument, crudely as follows: (A) Arguments are finite sequences of truth-value-susceptible expressions out of an object language, open to a variety of metalinguistic predicates, severally and sometimes jointly, such as "deductively (in)valid", "inductively (in)valid", "presumptively (in)valid" (for arguments from authority), etc. (B) As above, except that in place of the metalinguistic expressions we have corresponding object-language operators placed into the sentential sequences, taking sequences of sentences into sentences. Thus the operator "∴^{d}" would take ("Socrates is a man", "All men are mortal") into "Socrates is mortal"; "∴^{p}" might take ("This crow is black", ..., "this *n*th crow is black") into "All crows are black"; and "∴^{pre}" might take ("Zachary says he has a headache") into

"Zachary has a headache". On either approach, one is spared the necessity of ever saying that a bad argument is no argument at all, or that a good argument is good from every point of view (satisfies every metalinguistic predicate; permits blanket interchanges of its operator with other operators). Yet it is not obvious, certainly, that either view or argument compels, as Hamblin avers, the abandonment of techniques of formal logical analysis. True, only those B-type arguments of the form $\ulcorner \therefore^d ((A_1, \ldots, A_n), n+1 \urcorner$ can be said to fall within the competence of *deductive logic*, and if it is this that Hamblin means by "Formal Logic" it is entirely correct, and trivial, that Formal Logic is not the complete theory of argument.

3. DEDUCTIVE ARGUMENTS

Once she is liberated from her bad marriage to Formal Logic, the author has no further direct designs upon the divorcee. Indirection is counselled; rather than deal with the nature of arguments as such, the investigation turns to one of how "we" appraise and evaluate them. As part of a first approximation to a final account, Hamblin proposes that an argument is not good [80, p. 232] unless its premisses are true (and so doing indulges in another swipe at Formal Logic for permitting arguments that are falsely premissed yet deductively valid). But, for his own part, Hamblin also urges that sometimes there are good arguments for a given conclusion and equally good arguments as well for its negation. For that to be so, at least one of the arguments must be falsely premissd and *so not* a good argument after all. (On the other hand, the situation in which there are equally good argument *pro* and *con* is reasonably representable within formal (deductive) logic: an argument to A may be deductively valid; an argument to $\ulcorner -A \urcorner$ may be deductively valid; however, at most one can be sound.) That Hamblin has here sponsored more than a slip of the pen is pretty much unavoidable from his observation [80, p.233] that good arguments need not have true conclusions. Thus a good argument may proceed from true premisses to a false conclusion. No doubt, of course, there are from time to time arguments from true premisses to a false conclusion (e.g., that the earth is flat) such that when they were propounded they could not reasonably have been faulted or improved upon. Such arguments may well be in as good a logical shape as one could rightly have hoped for in the circumstances, but "as good as one could expect" is consistent with "bad".

Hamblin's first pass at a theory of good argument also requires that for an argument to be good its conclusion must, in some appropriate sense of "follow from", follow from the premisses *reasonably immediately* [80, p. 235]. In tracking the rationale of the requirement, it quickly becomes apparent that there is a confusion between a conclusion's following immediately from a set of premises and a conclusion's being, not only deducible, but actually deduced from a set of premises. As the author would have it, the first condition is met when the conclusion comes in just *one* step, perhaps even from just one premisse. And you have the second situation, if you have it not only that for the premisses G and conclusion A, $\ulcorner G \vdash A \urcorner$, but rather when you have D, where D is a deduction terminating in A such that the non-theorems of G are all members of D. Immediacy, understood the first way, is a silly and mistaken requirement, and understood the second way, it is a correct requirement, but it is silly to speak of "immediacy". What Hamblin wants, of course, is that his arguments be *deductions*; this is already provided for in the previous condition [80, p. 234] that there be a conclusion which in fact follows from the premises. Actually, on this very point the author discloses unsettled intuitions by a declaration, unadorned by rationale or explanation, (mindful of enthymetic set-ups?) that sometimes it is allowable that one's argument *not* be a deduction, viz., when premisses are permissibly omitted.

4. BELIEFS

What we have been calling Hamblin's "first pass" at a theory of good argument is a theory whose conditions are alethic. [80, p. 235] Hamblin believes, however, that a second pass, in which the alethic conditions are replaced by epistemic counterparts, makes for a general improvement in the theory. [80, p. 236] The new conditions provide that for an argument to be good its premises must be known to be true and that its conclusion must follow clearly from its premises; yet premises may be omitted when if unstated they would nonetheless be taken for granted.[1]

The second pass is possessed of troubled consequences. For one, it becomes clear that the theoretical objective is not the single predicate of arguments, "good" but rather the indefinitely large (denumerable) set of predicates $\ulcorner$good for v$\urcorner$, and so it is not even presumed to be within the theory's competence to distinguish *tout court* good argument from bad. That the key predicates of the theory are not less than dyadic prompts one to wonder whether there is any very good reason why they should not be *more* than dyadic. If a theory of argument is to deal seriously with such predicates as "good for Zachary", "not good for Pamela", "good for the Turks", why should it not be expected to extend its insights to predicates of greater polyadicity, to such predicates as "good for

Zachary while attentive," "good for person x when assuming y," "good for person x while in a state y at time t in place p under chemical influence c"; why indeed not for argument predicates of indefinite polyadicity (above monadicity)?[2] One may be forgiven for wondering whether the bias against the appropriate monadic predicates of arguments does not gravely imperil the whole enterprise.

It is clear that, under the second pass, the theory has shifted from its earlier focus on semantic properties of arguments to a preoccupation with beliefs, for when else is an argument valid *for* Zachary than when Zachary has the appropriate beliefs (in the appropriate way)? That this is so, is reinforced by Hamblin's additional epistemic condition to the effect that an argument is not good unless in its absence the conclusion would be in doubt. (So, roughly, an argument cannot be good for Zachary unless it at least raises, in Zachary's eyes, the credibility of its conclusion.) It is well to note that the condition once and for all settles, affirmatively, the question of the relativity of an argument's goodness or badness. Otherwise, our requirement (that no sentence A can be the conclusion of a good argument unless, sans its premisses, we have it that A is not known) requires that no proposition can be known save through the knowledge of those premisses; in particular, this seems to mean that if A is the conclusion of the good argument of which the premisses are B^1, ..., B^n, then A can be known only in one way (viz., via B^1 ... B^n), contrary to fact and the author's acknowledgement of it. There is also the point that, unless we adopt the relativistic stance, no proposition A of which I am uncertain can be the conclusion of a good argument directed to someone for whom A is uncertain. Thus, we must have it that the argument, bad for the arguer, is nevertheless good for the arguee, and Hamblin confirms this. [80, p. 242]

5. EPISTEMIC LOGIC

With the theory now pretty much expressly in quest of adequate things to say of such locutions as "believes valid" and "is in doubt as to whether" and so on, it is only natural to ask whether the natural terrain on which to conduct the hunt is an appropriate extension (or restriction) or adaptation of *epistemic logic*. The idea occurs to Hamblin, but it holds little appeal for him. [80, p. 238-39] For one thing he is concerned about the role of Superknower, but he soon betrays a misunderstanding of the rationale of Superknower. Superknower knows all logical truths because, classically, the logical truths follow from everything, and because knowledge is, in this theory, assumed to be closed under consequence. Superknower is an idealization designed to capture a rationality assumption. If Zachary knows that A and if B is a logical consequence of A, then, *in principle,* it is a measure of Zachary's rationality whether he can be gotten to realize that B. The arguments by which such realizations might be promoted need not be causally efficacious; it is enough that they "obligate" the recognition if only in the sense that non-recognition needs to be forgiven, excused, or at least explained away.

6. PROBABILISTIC MODELS

Persuaded that neither deductive logic nor epistemic logic satisfactorily captures (enough of) the structure of ⌜valid for v⌝, Hamblin turns his attention to a probabilistic counterpart of the earlier, abandoned, conditions. By these conditions in their probabilistic costume, a good argument would be required to have premises having a reasonable initial probability, a conclusion of which the initial probability is less than that of the premises, and an implication relation by which the probability of the conclusion is raised to match the probability of the premises. But Hamblin balks at the proposal [80, p. 240] because, since the negation of the conclusion implies the negation of the premises conjoined, the confrontation of premisses and conclusion could with equal justice be said to weaken rather than raise the probability of the premisses. Of course the same could be said *contra* the two previous sets of conditions - where, in the one case, the confrontation of premisses and conclusion modifies, *up or down*, the truth-values of the confronters and, in the other case, their epistemic value. But Hamblin makes it seem as though our current point does not apply in these cases, that it is peculiar to the probabilistic conditions. Possibly Hamblin is impressed by the following line of thought; under, say, the epistemic conditions, the appraiser of good argument is dealt two initial facts, namely that the premises are known, and, independently of the premises, the conclusion is not-known or neutral. Now (the argument continues) when we bring the two facts into mutual confrontation how *could* there result a reduction in the epistemic value of the premises, for do we not already have it that their value is "known"? And if they *are* known, nothing can reduce that value.

A bad argument, to be sure. If it were not, then it would count equally against our probabilistic model. If the probability of the premises *is* k, then it cannot be made other than it is. Of course, if we make a distinction between initial probability relative to a body of evidence, we can legitimize this talk of fluctuating probability-verdicts, but then there is no reason why a similar distinction shouldn't be made available for the vindication of fluctuating verdicts of knownness and truth. (Or must we deal with a strong and undeclared KK-hypothesis?)

We think, however, that Hamblin is right in thinking that neither the concepts of deductive logic nor those of probability theory will do for a satisfactory reconstruction of the structure of the goodness of argument and inference. In the simplest kind of case, where arguer and arguee are one and the same, in which we are concerned with soliloquial inference, a person has an initial stock G of beliefs which, in the light of some specified statement, A & G, he may have to modify. Inference may thus be likened to a function f from G to a set of beliefs D, where typically D = G ∪ {A} or D = G - {B}, where B is some member of G inconsistent with A. In the former case inference extends the initial belief set; in the latter it restricts it. What, then are the rules by which f can reasonably be thought to be governed? Not, as it happens, just or always by the rules of deductive logic.[3] For suppose that my initial belief stock G contains both A and ⌜A→B⌝; *modus ponens* prescribes just one procedure, viz., to extend G by addition of B. But, as Hamblin rightly observes (though only in a related connection), this is not always the good inference to make. Sometimes the thinking man must abandon ⌜A∧(A→B)⌝, thus restricting G, not enriching it. Similarly, if a man discovers his belief stock to be inconsistent, classical deductive logic, if it gave an adequate theory of inference, would oblige him to embrace the "universal" belief-set. But that is never the inference to make; rather our thinking man is required to modify his belief stock so as to eliminate the inconsistency. So the transformation rules of deductive logic are not the rules of inference.

The same may be said for the transformation rules of the standard probability calculus. Again, the really important insight is ready to hand, viz., that often the only good inference is to give up something previously held. The importance of the point can be shown in this argument due to Harman. [85, p. 92 ff.] Putting ⌜P(a, b)⌝ for the probability of *a* relative to b; and ⌜P(a∧b)⌝ for the initial probability of ⌜a∧b⌝; and ⌜P(b)⌝ for the initial probability of b, we have the law of the probability calculus:

$$P(a, b) = \frac{P(a \wedge b)}{P(b)}$$

But now, suppose that b is a part of my total evidence a; then b ⊨ a. So we have it that

$$P(a, b) = P(b),$$

and hence that

$$P(a, b) = \frac{P(a \wedge b)}{P(b)}$$

Yet if the correct inference to make is to *give up* a part b of my belief stock a, this flies in the face of the probability calculus by which the probability of a on b is unity! It is not therefore generally correct to say, given a "total evidence" belief set G and statement A, that as the probability of A relative to G approaches unity, the better is the inference from G to A.

7. DIALECTICAL PRECEPTS

It is not really all that surprising that Hamblin does not pursue his insight in the above ways, since by page 241 it finally becomes plain that his primary objective consists of developing a theory of good *dialectical* argument, in contradistinction to the usual sort of monolectical set-up one finds in books on deductive and inductive logic. It is Hamblin's contention that inasmuch as *dialogue* is the natural spawning ground of argument, only a careful description or reconstruction of the basic dialectical concepts holds much promise for a good theory of good argument. We have only to recall the author's earlier espousal of the thesis that a theory of argument cannot be good unless it provides a good solution to Mill's *petitio* charge to begin to savour the upcoming skirmish.

But what *are* the basic dialectical precepts? They include these [80, p. 242]:

(a) Dialectical give-and-take is paradigmatic of argument; arguing to oneself is at best the limiting case.
(b) In fact, it may well be that monolectical argument is "unintelligible" thanks to Wittgenstein's private language argument.
(c) To evaluate an argument alethically (i.e., in terms of truth and falsehood) is really only a special case of evaluating it dialectically (i.e., in terms of acceptability for him, acceptability for me, etc.) since in such a case arguer and arguee are one and the same.
(d) The alethic predicates 'true', 'valid', etc. differ in meaning in the first- and other-person cases.

(e) The dialectical "I accept. ...". is not equivalent to "... is true", but sometimes ⌜I accept⌝ is equivalent to ⌜A is true⌝. Namely, in the limiting case in which the addressee always agrees with the addressor. In such situations "A's argument in invalid" is synonymous with "I disapprove of A's argument".
(f) It is not, therefore, the pure logician's job or province to pass judgment on the goodness of any argument.

No case is really attempted in behalf of any of the items on this list. Hamblin seems to think that they are pretty much of that part of common knowledge from which attention has been deflected by formal (or pure) logic, and that having them recalled to our attention should suffice to command a renewed acceptance of them. Thus, we might agree that (a) is sensible; that (c) is interesting and probably worthy of careful development; that (f) admits of a correct interpretation but that it is not the author's; that (b) is silly and that (d) and (e) are hardly more than an expression of a crude semantic emotivism.

Still, there are any number of situations in which reasonable men barter one another's opinions for some ultimate gain (often a decision that such-and-such be done or an official affirmation that so-and-so is the case). The medium of exchange is acceptance and rejection, which need not and typically does not coincide with belief and disbelief, truth and falsehood, and knowledge and the lack of it. Such bartering is essentially argumentative (whether in the formal logician's sense or not is irrelevant), and it powers and informs a welter of corporate practices: bargaining, negotiating, pleading (as in law), and policy-making of all sorts. Irrespective of how badly the monolectical theories of inference and reasoning fail to capture the structure of such situations, it is interesting and legitimate to seek a statement of the conditions under which dialectical brokerage is sound.

Thus we have [80, p. 245] a fourth and final refurbishment of conditions previously met with. An argument is *good for v* (where *v* may be a person, a committee, the N.A.T.O. countries, and so on) only if: its premisses are accepted by *v*; the passage from premisses to conclusion must be of a kind accepted by *v*, unstated premisses must be of a kind as are *v* - acceptably omitted; and the conclusion would not be accepted by *v* save for the argument.[4]

This last condition is that by which Mill's challenge is presumed to be met. Our condition provides that if an argument's premisses don't increase the acceptability of the conclusion, it is not a good argument.[5] But there is a fatal ambiguity in "acceptability". Under one understanding, the logical truths cannot be conclusions of satisfactory arguments, given that necessity or apriority are marks of highest objective acceptability; yet in another, such proofs are paradigms of good arguments, (given that the theorems of mathematics and logic so often need their proofs in order to be credible or even to be understood). In a related understanding an argument is good only if it, as we may say, "enlightens" its conclusion, provokes the penny to drop concerning it. But in such an understanding no very useful contrast can be made between deductive and non-deductive arguments and so part of Mill's point is blunted. On the other hand, another aspect of Mill's attack seems largely to be vindicated. For can it seriously be proposed that *elementary* argument patterns of deductive logic (the various simple syllogisms, simplication, double negation, etc.) are patterns in which there is such a difference in initial acceptability between premisses and conclusion that their mere confrontation will alter the acceptability of the conclusion alone? Surely that could be so only for the logically dim, thanks to whom (it is disquieting to ponder) there might be a telling dialectical counter to Mill. Whether this apparent failure to deal with Mill *must* augur bleakly for a theory of dialectic is another and difficult question.

Chapter 2
Argumentum ad Verecundiam

1. ARGUMENTS FROM EXPERTISE

It is commonplace that the informal fallacies constitute a branch of logic (though we should prefer to say "the analysis of argument") virtually bereft of theoretical articulation. In recent years, the practical importance of the study of informal fallacies has been recognized in many an introductory logic text, but very little systematic or sustained attention has been given to exploiting the potentialities of this important area. Fallacies have been of relatively constant interest since Aristotle, but on the whole the tendency has been to approach them intuitively and taxonomically rather than to pursue their theoretical aspects at appropriate levels of generality and systematicity. One need not have done very much teaching of introductory logic to appreciate that the lack of clear and adequate characterizations of the fallacies makes it virtually impossible to commend to students fallacy-theoretic notions as a really effective strategy in the analysis of argumentation. Yet there can be little doubt that the usefulness of such strategies for adjudicating actual argumentation in natural language could be greatly advanced by a more systematic attempt to raise ourselves beyond mere intuitions about fallaciousness.

Consider, for example, the traditional fallacy of *ad verecundiam* or appeal to authority. It has often been urged that, like the other fallacies, the *ad verecundiam* does not admit of treatment by strictly formal methods, chiefly because of the psychological and contextual elements involved in its commission. De Morgan, among others, made a kindred point when he argued that since there is an indefinitely large number of ways in which one can make logical errors, we ought not to expect to hit upon any decisive method for determining when exactly a fallacy occurs [50, p. 237]. In the case of the *ad verecundiam*, whether or not to charge that a fallacy has been committed is often obscured by the plain fact that sometimes an appeal to authority is perfectly sound.[1] True, the appeal to authority might not be deductively valid, but all the same, it seems that on occasion it is wholly reasonable to give preferential credibility to the reasoned judgment of an expert over the judgment of a layman. Thus, it is often said that such an appeal commits a fallacy only where the appeal in question is to an authority that does not exist, to expertise of questionable or debased credentials.

But for all this talk of the occasional soundness of such appeals, it may be felt that there is always an element of intrinsic nonobjectivity in any appeal to authority and that objectivity is only to be found in a direct appeal to evidence, where whatever it is legitimate to mean by "evidence", evidence and expert testimony are understood to be disjoint.[2] By these lights, to allow an appeal to authority as a genuine form of acceptable argument is to throw scientific objectivity to the winds. How often do we hear it said that an explanation or prediction is "scientific" (i.e. reputable) only if it is intersubjective, reproducible, and so not dependent upon the private evaluation of a particular individual? According to this way of thinking an appeal to authority, having intrinsically inexact and subjective elements about it, must be ruled out of the domains of science entirely.

It is notable that in judgments of expertise the expert's verdict may be based on inarticulable background elements and so not be amenable to total sentential representation. It is sometimes said, therefore, that criteria of evidential support in terms of relative frequency or degree of confirmation relative to a class of sentences E, the evidence class, is not the sort of support set-up sought in the appeal to authority. And similarly, whereas explanation in terms of the deductive-nomological model suggested by Hempel requires that the antecedent conditions in general laws be stated in sentential form, in the appeal to authority this requirement is not always or even commonly satisfiable. In assigning a certain degree of credibility to a sentence, an authority may partly depend on his intuition or judgment formed from and enriched by substantial familiarity with the area in question. Such a factor is especially prominent in the case of the applied sciences, where diagnosis is not usually based entirely on an exactly specifiable evidence set. Yet here it is quite reasonable to depend on the reasoned valuation of an expert even though he is unable to cite evidence in the form of a specifiable set of sentences.

But surely it could be counterclaimed, an appeal to authority need not involve inexact and inarticulable elements. Consider the case in which amateurs are involved in a cardinality dispute. The one side puts forth the view that there can be only one infinite number, namely the number of all the integers. The other side demurs, citing the abundant reals. If in so doing the diagonal argument[3] were advanced, then the argument could be said to be settled by exact means and with all necessary articulation. On the other hand, if it were merely pointed out that Professor So-and-So, a mathematical whiz from Göttingen, had given assurances that such a proof exists, then we have an appeal to authority; but where is the intrinsic inexactness and where the inarticulability? The key question here is: Where is the element of authority? Professor So-and-So is an authority, and his expertise extends to the domain of enquiry to which he is privy, namely of the existence of the Cantorian diagonal proof, which is in fact by and large pretty much restricted to experts in his field. However, it is of some importance to see that in asserting the existence of such proofs, Professor So-and-So was not *exercising* his expertise. He was speaking from what might be called a special position to know, but it need not follow that claims made from special positions to know are claims made from expertise. A stranger who seeks, and receives, instructions from a native Montrealer as to the location of Jarry Park does not get an expert opinion, even if his benefactor is that city's Chief Cartographer. An expert's opinion need not be an expert opinion.

Thus, if we allow ourselves a distinction between *special position* claims and *claims from expertise*, it becomes not at all unreasonable to seek out aspects or elements of expert opinion which make it expert, not just the view of an expert. And at least this much seems clear: if a man, X, *takes* an expert's opinion as a judgement from expertise, *then as far as X is* concerned, there must be something about the case in question with which the expert is expertly familiar, but the exact articulation of which, if it exists, is unknown to X. Hence, if the expert were the cleverest physicist in all the world, and if he proposed proposition p and furnished for it a proof that no one else could follow (despite the fact that it was an exactly articulable proof for that expert), then the scientific community might nonetheless permit themselves the acceptance of p on the authority of that expert. If they did so decide, we would have a case in which the target of the objective evidence had to be foregone in favor of the weaker considerations of authority. From the expert's point of view it is as if he has spoken from a special position to know; but for the rest of us it is as if he were speaking from special expertise. Doubtless the distinctions we here seek are not nearly as well-behaved as one would like them to be. But certain gross patterns do seem to be discernible.

2. STOCHASTIC EXPLANATION

Debate as to whether forms of argument are genuinely scientific is not restricted to the fallacies. One has to think of the perennial debate between "hard" and "soft" approaches to science. Similar perturbations can be discerned within the metatheory of linguistics. Behavioral theorists are reluctant to admit appeals to the intuitions of native speakers, whereas Chomsky and others take the position that the resources of the so-called "hard" sciences can be partially utilized in less exact explanations without converting such explanations to those of "hard" science and *also* without impeaching the explanations for their failure to harden. It is a well-known conviction of Chomsky that data in the form of introspective judgments of the native speaker concerning well-formedness of sentences are actually what constitute the subject matter for linguistic theory. [36] To be sure, such introspective data are not sacrosanct; they may be challenged and evaluated by operational means familiar to the linguist. Yet operational methods, it is claimed, are not themselves always sufficient, and they may be overruled when they conflict with an introspective judgment.

Reluctance to accept Chomsky's account of the role of intuition in linguistics may stem from a kind of core positivism or behaviorism, in which it is usual to put a premium on deterministic systems and, correlatively, to eschew the loose, indeterministic explanations of so-called (and misnamed) mentalism. Yet it is increasingly recognized that deterministic explanations are not typical of the practice of science, and that the irreducibly stochastic explanations are characteristic of domains of enquiry having a high degree of empirical content. Not only are their concepts empirical and open-textured but relevant evidence may be available which cannot be completely specified or articulated now or ever; hence these judgments will exhibit a formal structure that gives them a predictive force even weaker than that of the more usual type of stochastic laws. Rescher and Helmer [163] argue that the usual deductive-nomological model of explanation must not only be modified to allow for stochastic laws but also to allow for what they call "quasi-laws". A quasi-law is a stochastic law that allows for exceptions by means of escape clauses that are tacitly understood yet not fully articulated. In such contexts, if an exception does occur, an explanation accounting for its exceptional character is available. According to Rescher and Helmer, it would be hopelessly impractical in these contexts to try to erect a theoretical structure similar to those in some more exact areas of science amenable to axiomatic treatments. In their view a properly pragmatic spirit, expressed in the desire for reasonable explanations and predictions, should encourage and endorse the use of quasi-laws. And if this should mean that the aesthetic niceties of a formal theory must be foregone where none is necessary or desirable, so much the worse for aesthetic niceties.

Thus, in linguistics we accommodate the concept of linguistic competence through appeals to native speakers, but because of the inarticulable nature of this competence the model of explanation is not of the usual deductive or stochastic varieties. Yet to discount such information for its not conforming to preferred scientific models of explanation is inconsistent with the actual practice of linguistics. We also find explanations of the quasi-law type in psychiatry, medicine, ecology, and elsewhere, particularly in the applied sciences; and who will say that such explanations are not scientific, objective, and rational? In the case of diagnostic decisions in medicine, or of political and economic decisions or command decisions in military affairs, it may be very important to make a reasoned determination on the basis of evidence that cannot be rendered exhaustively in sentential form and is not open to the usual data-processing techniques. According to Rescher and Helmer, in assessing the probability of an hypothesis, H, we are often required to rely not only upon a set of sentences, E, constituting the explicit evidence relevant to the hypothesis, but also on a body of background knowledge, K, which may be indefinite in content and not explicitly articulable in sentential form. This suggests the usefulness of turning to methods based upon a reasoned appeal to expert judgment rather than to methods which tolerate only the usual formal concept of confirmation.[4] Whether such a course convicts one of unscientific reliance upon the nonobjectivity of appeals to authority, or whether it recommends that our narrow, positivistic or behavioristic criteria of objectivity be expanded in order to make room for such forms or argument, it is a potentially fruitful line of enquiry.

3. THE LOGICAL STRUCTURE OF THE AD VERECUNDIAM

A preliminary statement of the form of the *ad verecundiam* argument has been provided by several authors. Rescher and Helmer say that an expert in some subject matter is a rational person who has a large background knowledge in that area and whose express or implicit predictions in his personal probabilities in that area show a record of comparative successes in the long run. As Rescher and Helmer point out, expertise involves a success factor which is relative to the predictive performance of which the average non-expert in the field would be capable. They cite the example that in a temperate climate a lay predictor can establish an excellent record by always forecasting good weather, but this would not support a claim to meteorological expertise [163, p. 34].

Hamblin proposes that the argument form

Everything X says is true.
X says that P.
Therefore, P.

is valid, and he observes that we could expect to find weaker, but still not fallacious, forms of argument within which some support is given to P by premisses of forms such as "X is an authority on facts of type so-and-so". Hamblin opines that *some* kind of calculus of testimony might be forthcoming and useful, however imponderable the factors in practical cases [80, p. 218].

Salmon suggests that appeal to authority has the following form:

X is a reliable authority concerning P.
X asserts P.
Therefore, P.

Salmon observes that this form is not deductively valid, but that it is, as he says, "inductively correct". It is, he believes, a special case of the statistical syllogism which could be rewritten as follows:

The vast majority of statements made by X concerning subject S are true.
P is a statement made by X concerning subject S.
Therefore, P is true [167, p. 64].

To set the stage for our evaluation of these suggestions regarding the logical form of the *ad verecundiam* argument, let us take a look at a number of considerations which typically should be weighed in adjudicating cases of appeal to authority. There are five main categories into which such considerations fall [167, p. 63-67]. It is our view that jointly they yield five conditions of material adequacy for any account of the argumentatively sound appeal to authority.

4. ADEQUACY CONDITIONS

AC 1. *The authority must be interpreted correctly.* In an appeal to authority common pitfalls are misquotation, mis-documentation, or inadequate rendering of the expert's expressed judgment. Subtle changes in emphasis in phrasing can be very important. If the claim is in the form of a direct quotation, it must be quoted in proper context. De Morgan warns us of the following pitfalls in the citing of sources (*Formal Logic*, 1847).

1. It is not uncommon, in disputation, to fall into the fallacy of making out conclusions for others by supplying missing premisses. One says that A is B; another will take for granted that he must believe B is C, and will therefore consider him as maintaining that A is C (281).
2. ... as to subjects in which men go in parties, it is not very uncommon to take one premiss from some individuals of a party, another from others, and to fix the logical conclusions of the two upon the whole party: when perhaps the conclusion is denied by all, some of whom deny the first premiss by affirming the second, while the rest deny the second by affirming the first (281).
3. ... quotation is [not] obligatory, though highly desirable: but the reader must remember, when there is only citation, that it is not the author cited who speaks, but the person who brings him forward... . If the citer be honest, the passage in question exists: if judicious, it is to the effect stated (282).
4. Perhaps the greatest and most dangerous vice of the day, in the matter of reference, is the practice of citing citations, and quoting quotations, as if they came from the original sources, instead of being only copies.
5. Unjustifiable as unnoted omissions may be, still more so are additions and alterations (284).
6. Omissions of context, preceding or following the quotation, may alter its character entirely: and this is one of the most frequent of the fallacies of reference, both intentional and unintentional (285).[5]

There is also the danger that an outdated expert opinion may be cited. If the expert has changed his mind, an earlier verdict loses much of its force. In other cases the appeal to authority may be ruinously misleading if certain qualifications added by the expert are omitted or suppressed. These factors are particularly important where the expert's probability-assessment is a conditional one relative to certain contingencies. If these contingencies are not preserved by the report of the expert's judgment, the appeal fails. If the authority's judgment is delivered in technical terminology, it may be required that it be rendered into a more readily intelligible form. This can be especially important where a panel of experts is attempting to make an interdisciplinary prediction. Injudicious technical jargon can have a disruptive influence in such cases, yet it is also to be expected that many of the strikingly fallacious forms of the appeal to authority trade on ambiguities, or vagueness, or misunderstanding in the translation of technical terminology to a usable language.

AC 2. *The authority must actually have special competence in an area and not simply glamor, prestige, or popularity.* Many of the appeals to authority typical of treatments in logic textbooks are cases which come under this heading. The difficulty of deciding when a putative expert actually has the required special competence is not easily overemphasized. There are many areas where it may be questionable whether there *are* any genuine authorities. There will be other areas where the means of evaluating the degree of expertise may be meager and only dimly understood. Even in those provinces where criteria of special competence may be relatively exact, there could yet be cases where a decision on rating the degree of expert competence is difficult. Some indices of competence may be suggested here, but obviously they are not to be thought of as standardized, nor is it very likely that an exact set of criteria will be forthcoming in the near future. For all its roughness, a provisional rating might be based on:
    i) Previous record of predictions, as suggested by Rescher and Helmer, and Salmon.
    ii) Where a previous record of predictions is not available, we may have a record of hypothetical predictions, that is, some sort of test that the purported expert may have undergone, that could indicate a rating of his degree of competence.
    iii) Access to a record of other sorts of qualifications in the form of degrees, professional qualifications, testimony of colleagues or other experts, or a record of familiarity with the area. The criteria, it is expected, will vary considerably from area to area.

AC 3. *The judgment of the authority must actually be within the special field of competence.* The most blatant of the fallacious appeals to authority occur where a legitimate expert in a given area makes a judgment that does not fall within it, but where added credibility is given to his judgment simply by virtue of his being an expert in *some* area. In many cases, particularly interdisciplinary cases, it may be difficult to judge whether or not a given sentence actually falls within a domain admitting or calling for some particular kind of expertise. Many areas of expertise in which the reasoned judgment of an expert is reasonably given a degree of credibility above that of a judgment of a layman do not have well-defined criteria of expertise.

AC 4. *Direct evidence must be available in principle.* It is reasonable to assume that for the appeal to authority to be adequate, the authority must base his judgment on actual, relevant, and objective evidence within the area concerned. This evidence need not be fully available; as we have previously said, it may not even be exactly specifiable. But where there is doubt, the authority must give some evidence that his judgment is objectively based. For example, if we have a panel of experts in a given domain of experience and the judgment of some authority on a certain question falls well outside the range of consensus, then in an evaluation of that particular judgment, evidence must be made available for the evaluation of the other experts involved. Similarly, where we have reason to believe that he may not be trustworthy for one reason or another, then an evaluation of the degree to which his judgment is based on relevant evidence would be called for.

But it will be objected that we are inconsistent to press the present condition. How, it may be asked, can one square it with the nonsentential aspect of an expert's inarticulable expertise that there must be objective supporting evidence? In answer, the intent of this adequacy condition is to mark off, and approve only one side of, a distinction between the judgment of an occultist who happens to enjoy a distinguished track record, and that of one who has genuinely expert knowledge.[6] Let us be clear in saying that we do not counsel against premisses of the form, "The ratio of successful predictions to (a *very very* large number of) predictions made by Madama Fortuna concerning the weather is, astonishing though it may seem, very nearly 100%". It is a simple matter to concoct natural sorts of contexts within which to embed such a premiss and, with the aid of "And Madama Fortuna predicts rain tomorrow", to conclude the wisdom of putting off the picnic until the weather improves. But in granting a measure of propriety to such an argument we do not offer it as an argument to *authority*. In fact, we need concede to Madama Fortuna no authority whatever in order to be moved by the abnormally high statistical correlation between her weather-wisdom and how the weather actually turns out to be. What we have here is not expertise (so far as one can tell) but a probabilistic connection whose statistical richness, though it amazes us and leaves us bereft of an explanation of it, is *there*, to be used, cautiously, and if need be.[7]

Appeals to authority focus squarely upon the expertise of the authority; they require an exercise of that expertise. As good an example as any of how this expertise may be exercised is this: any scientific theory worth its salt it under-determined by the relevant data;[8] hence all such theories essentially require the use of analytical hypotheses, which fill them out and bear much of their ontological weight. It is especially important to note that any given body of data admits, in principle, of equally good theorizations containing mutually incompatible analytical hypotheses. How, then, should conflicts among rival analytical hypotheses by adjudicated? Not, certainly, by appeal to direct evidence, by appeal to the data. One aspect of the adjudication procedure no doubt involves considerations of simplicity, coherence with neighboring theories, and the like. But the adjudication does not proceed definitively; there is always a kind of "leap of faith", for which we do, and are right to, turn to the theorist's mere *sense* of what is fitting and fruitful. Often this sense is an inexact and dimly perceived appreciation of, a currently inarticulable intuition[9] of, what in time may be articulably recognized as a body of doctrine on which the evidence *does* favorably bear. Objective evidence is always at hand, in principle, at least in the sense that the possibility can never be foreclosed that data will emerge which, jointly with other already acknowledged considerations, will occasion the *downfall* of the expert judgment. In a word, then, the demand that direct evidence be available in principle is tantamount to the requirement that the expert judgment have a certain semantic status, namely that it be in principle objectively testable.

An epistemological aspect of *ad verecundiam* arguments of methodological importance is that such arguments are said to be *opinion-based*. Dalkey writes that between *knowledge* and *speculation* there is a "broad area of material for which there is some basis for belief but that is not sufficiently confirmed to warrant being called *knowledge* [42, p. 2]. A common tendency is to lump everything not dignified by the title "*objective knowledge*" into the category of *speculation*, something devoutly to be avoided when engaged in scientific pursuits. It is this sort of dichotomy that we urge resistance against. While there may be an important sense in which opinion-based conclusions are "subjective", we would point out that such conclusions are very much a part of ongoing scientific practice, that they admit of scientific treatment and characterization, and that in no event are they simply to be equated with raw speculation or random guesswork. In fact, we see no compelling reason for supposing that a judgment from expertise can never be knowledge. Is all knowledge objective knowledge?

AC 5. *A consensus technique is required for adjudicating disagreements among equally qualified authorities.* An obvious problem with expert consultation is that authorities notoriously disagree. Yet since two (or more) heads are better than one, the rational method, when he is faced with disagreement, of one obtaining the required information (inconclusive though it may be) is to make a consensus. In fact, a *second* consensus after mutual inspection of the grounds of disagreement might even have a convergence effect. The usefulness of establishing a method for processes of consensus among experts has become increasingly apparent recently, so much so that considerable effort has been spent on the development of techniques for this purpose. Most notable have been the efforts of Norman Dalkey and others at the Rand Corporation to develop the so-called Delphi technique on which we will have more to say below.[10]

On the basis of the above consideration, it is suggested tentatively that the argument *ad verecundiam* may be said to have the following form:

1. X is a reliable authority in domain K.
2. p ε K.
3. X asserts that p.
4. p is coherent with relevant information obtained from other factors.
∴ p.

For simplicity, we shall occasionally speak of this proposal as the "analysis of the logical structure of the *ad verecundiam*", or, briefly, as the "analysis".

5. THE ANALYSIS ELABORATED

We will consider in order the premisses of the analysis.

(a) First, in discussing the concept of a reliable authority we must distinguish, though the distinction by no means *exhausts* argument from authority, between a *de facto* authority and a *de jure* authority. The realm of the *de jure* is performative: its modality is deontic or anyhow not straightforwardly alethic. A *de jure* authority is not an authority in virtue of being an expert in a certain area; his authority is invested in him in titular fashion, as with the clergyman who is authorized to perform the marriage ceremony. If *de jure* pronouncements are to be considered argumentatively, they may be said to have the following form:

1. X has the *de jure* authority in domain K.
2. p ε K.
3. X *declares that p.*
∴ p.

This form of argument appears to be deductive in the sense that it must be the case that given the premisses P1 to P3 the conclusion obtains; saying makes it so. The case of the performative saying so and making it so contrasts in interesting ways with the other kinds of "sayso" conditions.
  i) There is the case of the author who, in writing his novel, *seems* to make certain things true of his character. The analysis of the "sayso" situation in fiction is surprisingly complicated and will not be gone into here.[11]
  ii) A second kind of case is that of the token-reflexive in which, e.g., someone says, "I can utter an English sentence".
  iii) Yet a third sort of case is where the speaker seems to be in a best-evidence situation with respect to his own inner states, as when a man says that he has a headache. It is at this juncture we can begin to discern a clear line of passage from *de jure* authority to *de facto* authority. For in saying that I have a headache I do not bring it about that this is so; rather I report that it is so under circumstances of familiarity and access which require another's doubts about reliability to be bold ones. In fact, it may seem to some that in this kind of case we have the paradigm of the impeccable, though deductively invalid, appeal to authority. We shall return to this point below.

De facto arguments from authority are non-deductive yet alethic. It is interesting to observe the contrast between *de facto* and *de jure* appeals to authority because it is often in a confusion between the two that fallacious arguments arise. Historically it has not been uncommon, in giving sanction to decisions, to invest certain sources with infallibility and finality. Political and, in particular, religious authority has been used in this fashion to suppress heretical, disloyal, or radical opinions. It is important, therefore, to reiterate that in a *de facto* appeal to authority, a criterion of special alethic competence must be met. Another consideration, earlier discussed under adequacy condition 4, is that in principle evidence must be available to the expert. Thus, a genuine expert will be expected, to a reasonable extent, to present and explain the evidence on which his judgment was based.

The domain K in which the expert has special knowledge must be at least roughly delimitable. Many a fallacious appeal to authority is made when the domain K of the judgment appealed to is not an area in which we might reasonably expect an expert's word to be better than a layman's. A very contentious case is that of *moral* expertise [176]. In the *Meno*, Socrates asks whether there could ever be moral experts and whether virtue could be taught. We might reasonably question whether there can be such a thing as a moral authority, whether moral judgments of a magistrate, a priest, a politician, or a philosopher are to be given greater weight than those of a layman. In general, appeals to a moral or a valuative expert tend to be regarded as fallacious instances of the appeal to authority. Similarly, it might be debatable whether there can be genuine appeals to authority in matters of aesthetic discrimination. Of course, if morality is, as it is so often said, *as such*, incapable of supporting objective, reproducible judgments of *any* kind, so that there is an important sense in which such matters do not admit of the usual range of positive epistemic modalities (e.g. "known", "more reasonably believed than withheld", etc.) then they straightaway become irrelevant to our concern. We suspect that the skepticism with which the idea of moral expertise is met, at least the idea of *secular* moral expertise, is but a trivial reflection of this general skepticism about moral knowledge.

(b) Premiss P2 can be considered in the light of adequacy condition 3. A judgment p must actually be within the special field of competence. There will, in general, be nothing like a decision procedure here, and whether or not P2 obtains will be decided largely on the basis of how domain K is specified.

(c) As for P3, we may look back to adequacy condition 1. The authority must actually have asserted p at sometime. The report that the expert's opinion is that p must be interpreted correctly, as previously mentioned. P3 is violated when the assertion is wrenched from context, distorted, exaggerated, or falsely attributed to X, the authority. An interesting case here is where p has never actually been asserted by X but where X has asserted q, and q logically implies p. Another case is the one where X clearly would assert p, even if he has not at any time actually had occasion to do so. Such complications may counsel an amendment of

P3; and further research may indicate that P3 needs an endorsation-condition a good deal more complicated than that of mere assertion.

(d) Happily, P4 is somewhat more amenable to exact treatment than the other three premises. In fact, it appears to correspond fairly well to a known empirical technique, the so-called Delphi technique. The Delphi technique is a method for the systematic solicitation and collation of informed judgments on a topic. Delphi procedure is a set of sequential questionnaires interspersed with summarized information and opinion feedback derived from earlier responses. [185, p. 1] The purpose of the Delphi technique is to prevent professional status and high position from forcing judgments in certain directions, as frequently occurs when panels of experts meet. The intention is to ensure that changes in estimates reflect rational judgment rather than the influence of opinion leaders and trend setters [207].

In a Delphi, each member of a panel of experts in an area is asked to submit in writing his individual responses to questions; then the responses are made available to all respondents, and any responses that have fallen outside a certain range (often the interquartile range, the interval containing the middle 50% of responses) elicit requests for justification or evidence. Next, a further round of responses is collated in an attempt to cause convergence of opinions in the sense of reducing the spread of opinion. This technique is regarded not as a substitute for, but as an adjunct to, face-to-face panel discussion.[12]

According to Helmer, [89, p. 17] a convenient consensus formula applicable whenever the solicited judgments can be cast in numerical form (or linearly ordered) is to use the median. A variant of the simple median is a weighted median giving more than one vote to the opinions of experts whose judgments objectively deserve preferential treatment. Self-assigned competence scores may justify such differential weights. In addition to a consensus it may be desirable to have an indication of the spread of opinions among the experts; that is, we may state the interquartile range of their responses.

The proposed analysis is best regarded as a tentative suggestion that gives us a preliminary basis for organizing the factors to which weight must be given in deciding how to rate the degree of credibility of a particular appeal to authority. It would be overly sanguine to expect that such issues will be settled either definitively or soon. Yet this attempt to specify the logical form of the *de facto ad verecundiam* may be regarded as a stimulus to further research.

6. DIALECTIC

In appeals to authority our data are judgments delivered by an expert. These we evaluate in an attempt to obtain access to information to which we would otherwise have no access and to which he himself may have only an intuitive access. Thus the appeal to authority is essentially a two-person (or what we might call a *dialectic*) form of reasoning. For example, the Delphi technique involves at least two separate groups of individuals and at least four roles for these groups.

> First, we have the User Body: the individual or individuals expecting from the exercise some sort of product which is useful to their purposes.
>
> Second, we have the Design and Monitor Team (sometimes separate groups): the group which designs the initial questionnaire, summarizes the returns, and redesigns the follow-on questionnaires.
>
> Third, we have the Respondent Group: the group chosen to respond to the questionnaires. This may sometimes be the user body, or the user body may be a subset of the respondent group [207, p. 2].

The argument forms in logic and in models of explanation familiar to philosophers of science are what we might call monolectic. They are objective in the traditional sense, in that they do not involve essential reference to other persons within the compass or logical structure of the given explanation or argument; but in the area of logical fallacies, we have many arguments that are irreducibly dialectic, two-person, or even n-person, for manageably large n. In this respect, fallacies concerned with the concept of the burden of proof such as *ad ignorantiam* are especially notable. Where we have a debate or a dialectical situation between two persons, if X asserts p, it is quite reasonable for Y to respond: "What is your evidence or reason for asserting p?" X may wish to respond with evidence or may respond that he was not really asserting p, but was taking some other modal stance towards p, such as, say, assuming p or provisionally entertaining p. In a protracted debate (one thinks, for example, of Identity Theory disputes in philosophy) questions of the burden of proof may become very acute, very important for settling the outcome of the debate. The current neglect of such important considerations may be partly due to the fact that logic has evolved primarily on a monolectical basis, with the perhaps unintended consequence that the languages and techniques of standard logics are not adequate for basically dialectical situations. This may suggest an entire restructuring of the concept of argument. Instead of the monolectical arrangement of a set of premises and a conclusion, we might also consider models for arguments such as *ad verecundiam* that are essentially dialectical [80., ch. 7 & 8], [233].

It ought to be stressed that such opinion-based arguments as the appeal to authority are, though not illegitimate, no substitute for a direct appeal to evidence where such direct access is available. Reliance upon expertise is not a substitute for direct evidence, but it does very much appear that it is an indispensable adjunct to the reasonable and enlightened use of direct evidence. It is plausibly conjectured, of course, that as direct access to data becomes available the appeal to authority diminishes in respectability. On the other hand, with the increasing growth of specialized knowledge and the resulting decrease of universal access to what is known, it would seem dogmatic, to the point of truculence, not to recognize a proper role in argument for *ad verecundiam*. Having rid ourselves of a narrow, positivistic concept of the scope of scientific method, by which explanations perfectly typical of actual practice in the applied sciences are to be jettisoned, it becomes transparent that the systematic study of fallacies is an imperative for the theory of argument.

7. PRIVILEGED ACCESS AND EXPERTISE

We turn to a brief consideration of this question: "Is it or is it not the case that first person judgments about one's own inner experience constitute a paradigm of the non-fallacious appeal to authority?" We may say at once that adequacy conditions 1, 2 and 3 are clearly to be complied with by any good account of the acceptability of first person judgments of inner experience; that is, for one to rely upon such disclosures they should be correctly interpreted, the speaker must actually have the first-person competence in question, and his disclosure must not fall outside the special competence of the speaker. And it would also appear that normally condition 5 is met, for who shall compete with a man on questions of that man's firsthand experience of his inner life? But what shall we say of condition 4, of the condition that direct evidence must in principle be available?

There is a strong epistemological tradition according to which experience of one's inner states *is* direct evidence, in fact the best evidence one could ever want. If this is a correct view, then it seems that we must say that such cases are paradigmatic not of judgments from expertise but knowledge from a special position. On the other hand, there is another epistemological tradition according to which one's experience of one's inner states is so decisive as to make all talk of its being evidence grotesquely overstated. In neither event, therefore, does an appeal to a report of X's experience of his inner states seem to be an appeal to X's expertise. We say it seems so to us; but we are a good deal more confident in saying that these and related questions will benefit from further attention.

8. INFALLIBILITY

Finally, we conclude with the observation that the intuitive notion of authority, in a sense relevant to the *ad verecundiam*, does not seem wholly to be captured by the idea of the judgment from expertise. Although expertise may be said to be invested with a kind of authority, its authority is ultimately the authority of *knowledge*, with expertise serving as evidence of knowledgeability. But there is also in the concept of authority an aspect which is consistent with not the presence but the absence of knowledge, as when in the Common Law the existence of a precedent is given extraordinary weight. Understood one rather stark way, the Common Law regularly commits the howler of supposing that past decisions are, just so, good and binding decisions, and the greater howler yet, of supposing that a precedent decision that something is so implies indeed that it *is* so. Yet it is not our habit to judge the Law to be as intellectually bankrupt as all *that*, and it is useful to ponder why this is so. In its barest essentials the Law is unique among our practices and institutions for the degree to which it is permitted to serve as a secular model of infallibility. The Law creates its own truth, engenders its own judicial reality. And what is peculiar to the Law's knowledge of judicial reality is that it does not need to be determined by any independently verifiable fact about *natural* reality, that is to say, about reality. Presumably we think it is sometimes appropriate or anyhow not irrational to repose our trust in institutional mechanisms for the truths that they, as it were, create for us, not discover. If the Law is paramount or the most obvious among such mechanisms, it is hardly likely that truth-creation, as we might call it, is unique to it. It occurs to us that truth-creation is a factor in any serious theorizing and is also present among the sundry practicalities of everyday life. What this suggests is that in the not irrational acquiescence in such truth-creation we have an important parallel to the legitimate appeal to authority. Needless to say, the ultimate sanction, if there is one, for the metaphor of truth-creation is a deep and comprehensive epistemology.

Chapter 3
Petitio Principii

The fallacy variously called *petitio principii*, begging the question,[1] *circulus probandi*, and arguing in a circle, appeared on Aristotle's original list of fallacies, and it crops up in writings on informal logic in the treatises of the Middle Ages, through to De Morgan, Whately and Mill, and into twentieth-century introductory texts on logic. Its continued appearance testifies to its importance, yet, remarkably, the analysis of *petitio* has not advanced significantly since Aristotle's original treatment of it.[2] Our purpose in this paper is to attempt the beginnings of a systematic theoretical treatment of this fallacy.

That arguing in a circle is an essentially *informal* fallacy is easy to establish. Arguments of the form 'p, therefore p' always or nearly always beg the question,[3] yet their formal validity is impeccably reflected in standard first-order logic. Thus there is a kind of dissonance, markedly acute in the paradigm 'p, therefore p', that pervades all arguments in which the conclusion appears as a premiss or conjunct of a premiss. Such arguments are both valid and fallacious.

The truth is that *petitio*, like other informal fallacies, occupies something of a logical twilight zone. It appears rather too psychological for the austere tastes of mathematical logicians, who are properly wary of confusing mathematical logic with how people actually think; yet the *petitio* also appears a touch too formal and rigorous for the psychologist or epistemologist, who with recent tradition to back him, tends to think of it as properly an abstract matter of logic proper. We believe, however, and hope to provide some reason for others to think, that in a climate of such exclusiveness the deeper truths about *petitio* are scarcely likely to be found.

1. EQUIVALENCE AND DEPENDENCY CONCEPTIONS

In writings on *petitio* there are, broadly speaking, two main conceptions of the fallacy. According to the equivalence conception (as we shall say), an argument is said to be circular where the conclusion is tacitly or explicitly assumed as one of the premisses, that is, where the conclusion is equivalent, or even identical, to one of the premisses.[4]

> If one assumes as a premiss for his argument the very conclusion he intends to prove, the fallacy committed is that of petitio principii, or begging the question. If the proposition to be established is formulated in exactly the same words both as premiss and as conclusion, the mistake would be so glaring as to deceive no one. Often, however, two formulations can be sufficiently different to obscure the fact that one and the same proposition occurs both as premiss and conclusion.

The problem with explicating the equivalence conception is that while strict identity of premiss and conclusion is too narrow a criterion, capturing only the obvious cases, equivalence is too wide a criterion, attributing circularity to many arguments which plainly are not circular. De Morgan preferred the narrower criterion that, "strictly speaking, there is no formal *petitio principii* except when the very proposition to be proved, and not a mere synonym of it, is assumed" [50, p. 254]. However Sidgwick was of the view that nothing "appears to be really gained by restricting the name to so small a compass as this; and there is not doubt that such a restriction would be very much at variance with the popular acceptation of the term" [173, p. 194]. On the other hand, the wide criterion, requiring merely that a premiss be logically equivalent to the conclusion, would appear to condemn single-premissed arguments such as

> All rock-climbers are agile
> Therefore, no rock-climbers are not agile

to irredeemable circularity in every context. Sometimes it is said that an argument is circular where to state a premiss is to state the conclusion (in some suitably epistemically rich sense of 'state') [145, p. 26]. The trouble is, *does* one, in here stating the conclusion, in a suitably rich sense of 'state' also state the premiss?

According to what we shall call the dependency conception, an argument is circular where the conclusion is presupposed by a premiss or where some premiss actually rests on the conclusion, so that in order to accept the premiss, one need first accept the conclusion. Normally the 'flow of inference' in an argument is from the premiss to the conclusion

$$\left.\begin{array}{c}P\\C\end{array}\right\}$$

But where it is also required that an inference be made in the other direction, from the conclusion to the premiss,

$$\left(\begin{array}{c}P\\C\end{array}\right)$$

the argument iconized above is circular. According to this conception, an argument is non-circular only if one may know that each premiss is true without having to infer it from the conclusion, or from some other statement that can be known only by an inference from the conclusion [134, p. 34].

Thus there are two broad intuitive conceptions of circularity, each of which suggests various rough and by no means problem-free possibilities of explication as follows.

Dependency Conditions

(CD) The conclusion entails some premiss-conjunct.[5]
(CDE) In order to know that some premiss-conjunct is true, a must know that the conclusion is true.[6]
(CM) There is some premiss-conjunct that can be known to be true only by inference from the conclusion.[7]

Equivalence Conditions

(CQ) The conclusion is equivalent to some premiss-conjunct.
(CI) The conclusion is identical to some premiss-conjunct.[8]
(CQE) (For a) to know that a premiss-conjunct is true is (for a) to know that the conclusion is true, and vice versa.[9]
(CP) One has to state the conclusion in order to state some premiss-conjunct, and vice versa.[10]

Whether the two conceptions remain independent when more sharply formulated, or whether they collapse into one, or whether one is a special case of the other, remain to be seen.[11] Initially it may seem plausible that (1), the equivalence conception, is really just a special case of (2), the dependency conception. For, positing the conclusion as a (disguised) premiss is simply one way of introducing into the premisses a proposition that must depend on the conclusion because the proposition in question is that very conclusion itself. In the sequel, this plausible conjecture will tend to be confirmed by our examination of various circularity conditions. For the present, let us examine some elementary argument forms that pose interesting problems for the conditions.

2. THE DISJUNCTIVE SYLLOGISM

One of the first problems that our formulations of circularity conditions must cope with is that certain arguments, notably in the form of the disjunctive syllogism, are intuitively circular in some epistemic contexts but not in others.

Suppose, for example, one advances an inference in the form of the disjunctive syllogism

(1) p or q
(2) *not-p*
(3) ∴ q

in which the first premiss is construed as an intensional disjunction.[12] A proposition 'p or q' is put forward intensionally when one knows that one, at least, of p, q is so, but does not know which. Thus for the intensional 'or' it would be indefensible[13] for someone to utter 'p or q, and in fact p'. Assuming someone with rationality or logical acuity enough to make an inference in the form of the disjunctive syllogism, it is immediate that for the intensional 'or' this inference is indefensible for him to make. For, at the second premiss, he asserts the negation of p, and so commits himself to not-p; whereas at premiss one he commits himself to be non-committal with respect to p. There is no circularity here; but we do have indefensibility.

What this shows is that an inference is not always capably represented by a sequence of sentences relative to a fixed semantic and epistemic[14] background. In fact, inferences are frequently misunderstood unless they are taken *diachronically*. A simple setting for the disjunctive syllogism is as follows.

(1) I've long known that p or q, but not which.
(2) But I now discover that not-p.
(3) So, of course, that q.

Taken diachronically, the premisses of this (perfectly good) inference are assumed to possess temporal signatures which roughly mark the duration of their respective careers. When (2) is asserted, there are two logical verdicts, not one, available to the inference's assessor. One possibility is that since (1) and (2) are being co-asserted, the inference is indefensible. The other possibility is that (1) and (2) are not being co-asserted; that (2) requires and receives the retirement of (1) under its *intensional* interpretation, and its replacement some by suitably non-intensional rendering of 'or '.

Thus we have it that the premisses of an inference do not invariably commutate. 'p or q, not-p, $\therefore$ q' is not (virtually) equivalent to 'not-p, p or q, $\therefore$ q'. Neither do we have it that an inference '$p_1,..., p_n, \therefore q$' is good only if '$p_1,..., p_n$ only if q' is a self-sustaining sentence to utter. For, corresponding to 'p or q, not-p, $\therefore$ q', we have '((p or q) and not-p) only if q' which is equivalent to '(not-p and (p or q)) only if q', to which the inference might not in fact correspond. Yet one would have expected such a correspondence to be closed under equivalence.

If the intensional 'or' virtually implies the 'I don't know which' - disclaimer, there is reason to recognize a somewhat weaker sense of 'or' which, though it does not imply the disclaimer, is consistent with it. How does this, let us say *weakly* intensional, 'or' fare in the disjunctive syllogism? Assuming, as before rationality enough to make the inference to q, here too is an inference which it is indefensible to make in a non-diachronic setting. The point of conflict is that premiss (1) is consistent with the disclaimer 'I don't know', which itself cannot aide the second premiss, which asserts p's negation. Diachronically, however, the argument goes through in essentially the way it does when 'or' is taken in its strong intensional sense.

No well-brought-up student of sentence logic could long leave it unasked: what if 'or' is taken non-intensionally, i.e., neither intensionally nor weakly intensionally? Now if the force of the intensional 'p or q' is to the effect, 'p or q, and I know not which', and of the weakly intensional 'or' is to the effect 'p or q, and I need not know which ', the force of the non-intensional 'or' can scarcely fail to be to this effect: 'p or q, and I do know which'. If 'p or q' is rendered truth-functionally, then we have the significant *semantic* consequence that it owes its truth value only to the truth values of p and q. If 'p or q' is taken non-intensionally, we have the significant *epistemic* consequence that knowledge of 'p or q' owes itself to only the knowledge of p and q. This epistemic fact is a precise analogue of the semantic fact; that is, non-intensionality is an epistemic counterpart of truth functionality.

Assuming again rationality enough to complete the inference to q, we can safely allow that anyone competent to make this inference will recognize, and be bound accordingly by, the incompatibility between p and not-p. Such a person, therefore, cannot know that p or q without knowing that q. But since q is the argument's intended conclusion, it commits a classical *petitio*.[15]

There is an instructive lesson in this. It is the very lesson that Anderson and Belnap have been trying to teach us about *entailment*. But, as it happens, it is better understood as a lesson about inference,[16] a lesson about an epistemic, not semantic, impropriety. Loosely put, it is that semantic virtues may, in their epistemic analogues, be unredeemable rogues. It is a deep lesson. No deductive inference that may be cast in the form of the disjunctive syllogism (or its equivalent *modus ponens*) - let us not here worry whether these are all or just nearly all such inferences - may contain the non-intensional 'or' as main connective of the major premiss. The epistemic counterparts of truth-functionality are ruinous for *inference*. They commit all such inferences to the Carousel; and after a fashion they vindicate - completely or nearly so, it hardly matters - poor Mill. We need only remind ourselves that the premisses of the disjunctive syllogism jointly have the form '(p or q) and not-p', which is equivalent to 'not-p and q'. Thus our inference is in the form

 not-p and q
 ∴ q

which is circular by our condition (CI). And of course *any* argument castable in the form *modus ponens* or its equivalent is exposed to the same, abrupt fate.

It is clear why semantic virtues may be epistemic liabilities. The semantical property of truth-functionality is a virtue for entailment; its epistemic analogue, non-intensionality, is disastrous for inference. It is hard to imagine how, for example, entailment could get by without the law of simplification, '(p and q) only if p'; but precisely that which inference cannot get by *with* is the counterpart of that law, namely 'p and q, *therefore p*'. What Mill may have been struggling clearly to perceive was that the inference-theoretic counterpart of what we know as truth-functionality just is circularity - provided the connectives are taken non-intensionally. The moral, of course, is that inference does well to steer clear of the epistemic analogues of truth-functionality.

The disjunctive syllogism tilts instructively against truth-functionality under epistemic mutation; it shows the importance of not supposing premissory order to be commutative; it suggests the need to recognize that not every premiss is retained throughout the career of an inference - that, indeed, earlier premisses are often retired by later ones; and it also readies one to be receptive to more general truths, for example, that inference, all inference, is deeply sensitive to the epistemic background against which it chances to find itself prosecuted. This sensitivity to variously shifting epistemic backgrounds is well-remarked by Archbishop Whately.

It is not possible ... to draw a precise line, generally, between this Fallacy [petitio] and fair argument; since, that might be fair reasoning, which would be, to another, "begging the question;" inasmuch as, to the one, the Premises might be more evident than the Conclusion; while, by the other, it would not be admitted, except as a consequence of the admission of the Conclusion [208, p. 179].

Recognition of such variation in contexts of the disjunctive syllogism does not, however, commit us to a wholesale subjectivism concerning circularity.[17] For, if two persons to whom an argument is directed both agree in their appreciation and rating of the epistemic and evidential circumstances of the argument, presumably they could also come to agree on whether or not it was circular. What such variation *does* indicate is the usefulness of a move from non-epistemic conditions like (CD) to richer conceptions of dependency-circularity such as (CDE) and (CM).

It can also be argued, quite generally, that the expression 'deductive inference' is a misnomer. Entailment is one thing; inference another. It is the theory of entailment that charts the waters of formal deduction; and its instruments of navigation are satisfiability, validity, semantic entailment and the like. Inference however is not adequately represented except in an epistemic idiom. And it is well-known that satisfiability, validity and semantic entailment are notions that do not suffice for the important truths about epistemic discourse. What is needed are concepts such as defensibility, self-sustenance and virtual implication. These are notions with which to make some theoretical headway with inference; but they are not formal *deductive* notions.

Perhaps there is also point in urging a reform of the word 'argument' which presently does double-duty (at least), now for the semantic notion of a (valid) proof, and then for the epistemic notion of a (sound) *inference*. Here is an ambiguity of moment enough to map important truths onto palpable and incredible falsehoods; an ambiguity, therefore, not to be tolerated.

3. IMMEDIATE INFERENCE

Problems for the equivalence conditions are posed by the occasional non-circularity of immediate (one premiss) inferences. Under some conceivable circumstances, for example when one holds that all rock-climbers are agile, but denies that no rock-climbers are not agile, it may be non-circular to say:

All rock-climbers are agile.
Therefore, no rock-climbers are not agile.

Non-circular, yes; but it is not transparent that this is an *inference*. It sounds better as a statement of the very equivalence of which the person seems unaware. It is a piece of information for him, a remedy for his ignorance, not for his illogicality. Indeed it is hard to see how there could be such a thing as circular *inference*. Inference is a kind of belief modification, which can be represented by a function from sets of beliefs to sets of beliefs. The sets which are in the range of this function may be either supersets or subsets of, or just different from the sets in its domain, depending upon whether a belief is merely added, or merely subtracted, or an old belief is subtracted and an new belief lodged in its place.[18] Whatever might be the rules which govern the belief-modification function, it is not plausible to think that they should proscribe one's coming to believe anything that happens to be entailed by what one presently (and consistently) believes, provided it is properly a candidate for a *new* belief. True, there could be something strange, perhaps irrational, in someone's thinking that he now believes what formerly he didn't when, in fact, he formerly did. This is especially acute when the 'old' belief is set out as a premiss, the 'new' belief as conclusion, and the 'old' and 'new' are transparently identical. Thus the circularity, as we have been calling it, of an inference in the form of the disjunctive syllogism, under the non-intensional interpretation of 'or', comes basically to this: that the inferer fails to note that his 'new' belief is *not* new, that it is the very belief on which his major premiss rests. The crime, if there is one, is less one of logic than one of psychology. This is not fallacy, this is dissociation.

Whether an immediate inference betrays this deficiency must surely depend upon particular circumstances. Does a person disgrace himself by admitting that once he had not realized that 'All A are B' yields up 'No A are not B', that it was not until, as an undergraduate, he started thinking about quantifiers that he made this inference? No doubt part of the difficulty of appreciating in a general and a priori way the extent of a man's future beliefs that may, relative to his present beliefs, only aberrantly be called 'new', is reflected in the difficulty of knowing in a general and apriori way the extent to which belief is closed under consequence and equivalence. It is also, in part, reflected in the difficulty of determining just when 'a believes p' should be expected to give rise to 'a believes that he believes that p' or to 'a

knows that he believes that *p*'. If someone believes that the apple is red, it is to be expected that he will believe that the apple is coloured. If he comes to infer that the apple is coloured, then his domain belief set contained the belief that it was red, but not the belief that it was coloured. So merely that he comes to *infer* that it was coloured may show that his previous belief state was aberrant. That he comes to infer that it was coloured, *immediately*, without the aid of any heretofore unavailable premiss, may just show that he has snapped out of his doxastic inattentiveness; it may also give him the opportunity to say, and to be taken seriously in so saying, 'I hadn't realized that I *did* believe it to be coloured'. Possibly nothing very promising can be said of the immediate inference until we can be more confident than we presently are about the identity conditions of the objects of belief. Pending that achievement, it nonetheless seems clear that the immediate *inference* provides a rather questionable vehicle for circularity.

We earlier remarked - or anyhow claimed - an ambiguity in the word 'argument'. It may be instructive to return to the point. There is indeed some truth in saying that 'argument' competes for two quite distinct notions, proof and inference. But our reflections, just now, upon immediate inference, encourage the idea that there is, perhaps, yet a distinct third concept on which 'argument' makes the best possible claim, the concept, namely, of *an argument*! If proof is taken in a broadly axiomatic way, then that a proof is circular or question-begging cannot be complained of. The proof which is a one-membered sequence of an axiom is a proof of that axiom; it is not any the less a proof for being in the form

$$\langle \emptyset \rangle \quad \text{Q.E.D.}$$

That is just the way of proofs.

An argument, on the other hand, is in its basic form, a *claim* upon one to modify one's belief-state. It is a dialectical challenge to a person's consistency; it is a challenge in the form 'But that is indefensible for you to utter' or 'Since you grant this and that and the other, it is unreasonable of you not also to grant such and so'. Whether an argument succeeds does not even turn upon whether he to whom it is directed does *in fact* believe such and so (he may or may not) but upon whether his not believing it is indefensible for him.

Arguments are open to unique failures, not least of which is by the challenge to believe a given proposition, *p*, on the basis of beliefs which the challangee[19] does not in fact possess. In the paradigm '*p*, so *p*', circularity is just this particular failure writ large, and grossly manifest. In the, as we shall now call it, immediate *argument*, epistemic context again plays an essential role when it comes to the question of circularity. One such context may be schematized as follows;

A: Not-q
B: But you believe that p, do you not?
A: Ah yes, I 'd forgotten; of course, q.

Intuitively, there is no circle here; just a memory jogged. There is however the important task, granted that belief implies neither believed-belief nor known-belief, of determining the deficiency, if any, of failing to recall one's belief in p even in the light of the explicit rejection of q, its immediate implicans. It would appear, again, that the threat is not so much B's circularity as it is A's epistemic dissociation.

A second epistemic context is worthy of note:

A: Not-q
B: But you believe that p, do you not?
A: Indeed
B: You mean you didn't know that p implies that q?
A: No I hadn't realized this; very well, q.

Let us note at once that this is not an *immediate* argument, that its gross form is not '$p \therefore q$', but 'p, p only if q, $\therefore q$'; and that the premiss 'p only if q' seems to be accepted by A on B's sayso. So the argument may involve an aspect of the *ad verecundiam*.[20] Possibly some readers will think that A is involved in something like the reverse of Achilles' role in his classic debate with the Tortoise [34], [cf. 219]. Achilles was happy to concede propositions in the form 'p, and p only if q'; but he declined to see why he could not consistently refuse the affirmation of q. A's situation is one in which he abandons 'p, but not q' just on B's assurance of 'p only if q'. If Achilles' error was blind intransigence, it may be that A's is blind acceptance. But of course it need not be that. Perhaps the more urgent question about A, granted that belief is not closed under consequence, is whether there is any failure, for this particular p and q, in not believing q, even though p is believed. One might think so if p were 'The apple is red' and q 'The apple is coloured', but whatever the precise nature of A's deficiency, it seems entirely a separate matter from argumental circularity. True, for A to accept the premiss 'p only if q', the argument '$p \therefore q$' being *immediate*, is tantamount to his accepting '$p \therefore q$' is a good argument', which seems to beg the question against A. But it needn't do this, any more than is one's being convincedly corrected in one of one's assertions ('*p*'; 'No, not-*p*'; 'Oh, I see, I didn't know').

For a logically omniscient being (Hintikka's Superknower) all arguments would be circular in which the premiss is equivalent to the conclusion. For the rest of us, whether or not such arguments are circular depends on how transparent the equivalences are. Nevertheless if our ability standardly to perceive such equivalences were significantly a function of the complexity of the argument, it might be possible to give an objective account of 'standard circularity' insofar as we can introduce measures of complexity. In a different context, Hintikka has suggested such measures of 'information content' for first-order languages based on the number of interdependent variables in sequences of nested quantifiers.[21] It is quite conceivable that similar measures could be devised to capture circularity for a person of a given degree of logical acuteness. But we do not pursue this question here.

Finally, there is an additional problem related to immediate argument that tells against (CQ) and is impervious even to an acute-reasoner qualification. The following argument appears, under certain circumstances, to be non-circular.

> If Alvin is missing, so is Calvin, and Alvin *is* missing.
> Therefore, both Alvin and Calvin are missing.

The premiss is equivalent to the conclusion, and thus, by (CQ), the argument is circular. But even if directed to a logical sophisticate, the argument might be non-circular. Consider the case where Alvin and Calvin, two alpinists, have departed together on the expedition to Mt. Robson, and it is determined that Alvin is missing. We can easily imagine such a context, and relative to it, the argument above seems non-circular. One way might be to insist that the 'premiss' is really *two* premisses so that no one premiss is equivalent to the conclusion. In other words, we could add the restriction to this formulation that all premisses must be conjunctively atomic. Alternatively, we could say that the above 'argument' is really *two* arguments, i.e., we could add the restriction that the conclusion of a non-circular argument needs to be conjunctively atomic. Yet what seems to be merging as a best solution is simply the abandonment of (CQ).

4. UNIVERSAL ARGUMENTS

Another form of argument provides some interesting problems for our circularity-conditions, namely *universal arguments* - arguments that purport to claim that some particular A is B by bringing forward the premiss 'Every A is B'.[22] Suppose we are asked to prove that this bicycle is Hector's and that in reply we argue

> Every bicycle in the neighborhood belongs to Hector.
> Therefore, this bicycle belongs to Hector.

We might be justifiably accused of begging the question. Or suppose we were to defend the thesis that Hermione is thrifty by bringing forward the argument,

> Hermione has all the domestic virtues.
> Therefore Hermione is thrifty.[23]

Have we begged the question? As with the previous inferences and arguments, it seems quite conceivable that there are evidential and epistemic circumstances in which these two arguments might be non-question-begging; but intuitively, as they stand, they blatantly commit the fallacy of *petitio principii*.

The interesting thing is that neither argument is circular by any straight-forward application of (CD), (CQ) or (CI). Consider the bicycle argument. The conclusion does not entail nor is it equivalent to a premiss-conjunct unless, of course, we read the premiss as a kind of disguised conjunction, 'Bicycle$_1$ in the neighborhood belongs to Hector and Bicycle$_2$ in the neighborhood belongs to Hector &...& Bicycle$_j$ in the neighborhood belongs to Hector'. Thus the conclusion, 'Bicycle$_i$ in the neighborhood belongs to Hector' will be identical with some premiss. But such reflects a dubious view of the quantifiers.[24]

How do the epistemic conditions cope with arguments like the bicycle-argument? (CP) seems not at all appropriate since it is not clear that to state that 'Hermione has all the domestic virtues' is to state 'Hermione is thrifty' in any sense of 'state' that would not make virtually any deductively valid argument circular. (CQE) seems to work no better than (CD), (CQ) or (CI). This leaves (CDE) and (CM). How well these conditions cope with the circularity of the bicycle-argument is best seen by reflecting on what seems intuitively remiss about it. Principally there are two features of it that occasion complaint. (1) In such an argument, one serious difficulty may be that the premiss is advanced without any, or sufficient, evidence. But *bereftness of evidence* is not circularity, even though a circular argument may suffer from it as well. This failure of evidence is especially acute when, (2), the person to whom the argument is directed believes that the object described by the conclusion is precisely a *counter-example* to the generalization which is the premiss. True, the arguee's belief that the object of the conclusion is a counterexample is a deficiency even if the premiss is supported by reasonably strong, though simple, inductive evidence. But where the two deficiencies co-occur in an argument, circularity seems an inevitable result.

(CDE) and (CM) appear well enough to tally with this explanation. According to (1) and (2), the argument fails because, given the lack of independent evidence for the premiss, the disacceptance of the conclusion virtually implies disacceptance of the premiss. Hence the argument is circular by (CDE) since for the arguee to acknowledge that the premiss is true, it would make it indefensible to withhold the conclusion. It is worth nothing that such might not be the case if we had evidence independent of the conclusion that the premiss were true. In that case the argument might evade the Carousel.

For all its apparent similarity to (CDE), (CM) does not ring quite true in these contexts. It is not obvious that the premiss of the bicycle-argument can be known to be true only *by inference* from the conclusion. In fact, more is required for the truth of the premiss than merely the truth of the conclusion, and it is somewhat misleading to say, following (CM), that there need be an inference from the conclusion to the premiss. But provided we understand (CM) in a weakened fashion, such as

(CMI) There is some premiss-conjunct that can be known to be true only by inference from the conclusion in conjunction with a (possibly empty) set of additional statements (with the exception that the premiss-conjunct must not be known to be true by inference from these other statements alone).

This conclusion seems as applicable to the bicycle-argument as (CDE).

Even so, yet another example will enucleate a significant difference between (CDE) and (CMI). G.E. Moore, in the famous 'Proof of an External World' [139] offered to 'prove' the statement 'Two human hands exist' by furnishing premisses as follows: Moore says, "Here is one hand" as he makes a certain gesture with his right hand, and "Here is another" as he gestures with the left. Imagine, however, that one were to advance a somewhat different argument for the same conclusion:

(1) Everyone in the room has two hands
(2) ∴ Two human hands exist.

Note that by (CMI) the argument is not circular; for plainly it is possible to know the premiss to be true other than by *inference* from the conclusion. Even if the setting of our argument is such that one could not know the premiss without already knowing the conclusion, it hardly seems probable that the universal premiss would be inferred from it, or could. True in such a setting, the argument is circular by (CDE), but not we think by (CMI).

On balance, we would suggest that the burden of evidence speaks more favourably for the criterion of circularity (CDE) than for (CM) and its variants. The superiority of the epistemic criteria, (CDE) and (CQE), reflects a, by now familiar, feature of inference and argument: it is that they exhibit significant variation from epistemic setting to epistemic setting.

5. CIRCULAR DEFINITION

Here is a simple example form Sidgwick: Every effect must have a cause, since otherwise it would not be an effect. This draws attention to the role definitions often play in circular arguments, though we don 't say that *this* definition is circular. We have little to say here about the notion of circular definition,[25] since our primary concern is with circular argument.[26] But the Sidgwick example shows that definition and circular argument are often connected. What grounds does the premiss of this argument offer for the conclusion that every effect must have a cause? "It is defined that way". This response may beg the question of whether the definition is adequate,[27] although it does make the possibly informative additional claim that the conclusion is meant to be analytic truth.[28] Many examples of *petitio* cited by logic texts are of this type.

Another important area where circular argument and definition intersect is that of question-begging-susceptible appellatives as in

> (D) This doctrine is heresy.
> Therefore, this doctrine should be condemned.

Perhaps the most thorough and astute account of this type of fallacy is that in Bentham's *Book of Fallacies*, Part IV, Ch. I. [19]

6. EFFECTIVENESS AND SOUNDNESS

We shall now attempt to deepen our understanding of *petitio* by briefly clarifying some general properties of arguments in a way that will enable us better to articulate our intuition that *petitio* is somehow a failure of an argument's effectiveness or persuadibility rather than a failure of its soundness or validity.[29] Let us say that *q profortifies p* where the probability of *p* given *q* is greater than one-half (*q* makes *p* more likely than not). Let us also say that *q befortifies p* where the probability of *p* given *q* is greater than the initial probability of *p* (*q* makes *p* more likely than before). Third, let us say that *q fortifies p* where both of the previous relations obtain, that is, where *q* makes *p* more likely than not *and* more likely than before. Finally, we might say that *q weakly fortifies p* where neither *q* profortifies *p* or *q* befortifies *p*. If we include the deductive case where *q* entails *p*, we will have a total of five ways in which a set of premises may be argumentatively related to a conclusion. For this reason we call these five relations the *relational factors* in an argument. In a sound argument, in addition to the relational factor, the initial degree of support of the premises is also important - we might call this the *initial factor*. Thus we will here say that an argument is *sound* where one or more of the five relational factors is met, and moreover, the premises have a certain degree of probability, k. We have explored some consequences of this view of 'sound argument' elsewhere, [221] and have there shown that there are various kinds of sound arguments of these types of which it is true to say that '$p \therefore p$' is a sound argument. That *circulus probandi* is not a fault of sound argument in this purely alethic sense is hardly surprising, knowing, as we do, that circularity is largely an epistemic matter pertaining to 'epistemic settings' and 'logical acuity'. Circularity is not the failure of an argument to be sound, but rather a systematic failure of an argument to be effective. A circular argument is fallacious because it systematically lacks the power to escalate the credence of its conclusion for those to whom it is directed.

The study of fallacies is at once psychological and logical. Effectiveness is its psychological component, soundness its logical component. Neither is sufficient for an argument to be adequate (in a broad but common sense of 'adequate'). Both components are required if we are to develop a theory of argument worthy of the fallacies.[30]

The notion of soundness, as we view it, pertains to the judgment of an abstraction, namely an ideally rational reasoner. For current purposes, soundness is relative to whatever the reader considers the appropriate account of deductive and inductive logic. Perhaps it might be relative to some system of first-order logic and some axiomatization of the probability calculus. In contrast, we here suggest a notion of *credence* that exemplifies the actual beliefs of a flesh-and-blood individual. The base notion of credence incorporates no ideal rationality assumptions. One example of such freedom form ideal rationality assumptions would be that the concept of credence does not require that

$$(B_a(p) \ \& \ B_a(p \rightarrow q)) \rightarrow B_a(q)$$

be a theorem. This notion of belief or credence refers to the personal belief system possessed by an actual individual. Let us assume that the strength of such a person's credence with respect to a belief of his, Cred (p), can be assigned a real number ranging from 0 to 1, and similarly for conditional credence Cred (p, q) [38]. Admittedly, the logic of credence as we might loosely call it, is not well-understood, nor might it turn out to be generally possible to attach numerical values to credence in this way; but for simplicity and clarity let us proceed as if it were, tentatively.

Pr(p), the probability of p, represents an objective notion - numbers are assigned to p in virtue of underlying stochastic data-processing procedures of the kinds familiar to statisticians. Cred (p) represents a subjective or personal notion - it is the strength of an actual person's belief in p, estimated behaviourally or attitudinally by techniques of psychological measurement. Pr(p), roughly speaking, represents the strength of belief of a perfectly rational person (an idealization) who understands the relevant statistics. Cred (p) represents the strength of belief of a given flesh-and-blood actual person, who will be imperfectly rational.

We concede that the notions of effectiveness and soundness of argument adequate to fallacy-theoretic contexts are not presently understood at any very high level of theoretical generality, and that the tentative explications advanced above are little better than crude beginnings. But they are badly needed beginnings. And assuming that they have some degree of initial plausibility, at least in outline, we can gain some theoretical understanding of the notion of *fallacy* through them. A fallacy occurs where there is a pattern of wide disparity or skew between effectiveness and soundness in a class of arguments. Now an individual discrepancy is simply a blunder, mistake or error and not a fallacy. But where a family of such discrepancies is traceable to some isolable general cause, then something we may label as a fallacy may emerge. A fallacy is a recurring discrepancy between effectiveness and soundness that can be pinned down to some cause. A fallacy is a snare and a delusion, a particularly common type of pitfall, an insidiously persuasive pattern of argument, a virus that can be isolated, identified, tracked down, and inoculated against.

If an argument does not raise the credence of its conclusion for a then it is ineffective for a but not necessarily fallacious for a. But if the argument is such that it could not possibly be effective for a because the credence of the premisses is bound to the credence of the conclusion, then that argument is doomed to be ineffective for a and commits the fallacy of *petitio principii*. But how can this condition be met, making no ideal rationality assumptions? If it is not true that some persons would believe anything, it is certainly not obviously false that there is nothing which someone or other would not believe. We concede in someone certainly the theoretical possibility of a gullibility so vast as to make the rest of us forever secure from committing *petitio* against him. But it should be emphasized that, so far, *petitio* is an individual matter. What commits *petitio* for you may not do so for me. Yet insofar as factors such as 'epistemic setting' and 'logical acuity' can be *standardized*, can be estimated to be equal in a set of individuals, thus far can an argument be said 'objectively' to beg the question. What sorts of modal profiles such 'standardized impossibilities' might conjure up is a difficult and interesting question which we do not here pursue. Rather we would commend these sundry matters to the interdisciplinary attention of a logically mature psychology and a psychologically aware logic.

Chapter 4
Ad Baculum

The traditional area of the informal fallacies is a much understudied domain. The importance of the fallacies can be recognized by their constant discussion in logic texts since Aristotle, but *The Standard Treatment* [80] of fallacies, with its hackneyed examples, and bereft of theory, is an embarrassment. Small wonder that logic has fled to the more invigorating climate of mathematical articulation. The fallacies have been left behind, while other departments of logic, to their immense benefit, have been merged with the more formal disciplines.

In the hopes of helping to initiate a reversal of this tradition of casual negligence, we here propose to examine the fallacy of *ad baculum*, the appeal to force. Most texts treating of the fallacies include a section on the *ad baculum*, in which it is lamented that we should so often turn to the sword instead of the pen as a means of persuasion. Our concern will be whether such threats can be said to constitute instances of a *logical fallacy*. The weight of evidence persuades us that the answer is, "No".[1]

1. THE STATEMENT REQUIREMENT

A persistent annoyance is that most text-book accounts of the *ad baculum* cite amongst its instances "arguments" that turn out not to be arguments at all. Yet is seems reasonable to require that an informal fallacy should be an incorrect *argument* and not simply a falsehood, a silly belief or practice, a behavioural aberration, a breach of good taste, or some other lapse from civility. Of course, it is not transparently easy to specify what should count as an *argument*. Elsewhere we have suggested that one needs to take a fairly liberal view of argument if one is to deal effectively with the fallacies, since narrowly alethic or classical proof-theoretical representations of argument are not nearly rich enough to capture the required nuances. Many fallacies are in fact essentially epistemic or dialectical. But it would be ungeometrical if not inflationary not to draw the line somewhere. A simple ruling that would preserve many of the intuitions underlying the treatments of the fallacies to be found in logic texts, and yet be consistent with the more well-developed branches of logic, would be as follows: an argument is an ordered pair of sets of statements - a unary set called the conclusion, and a finite non-empty set called the premisses. Statements, in turn, are required to admit of truth and falsity. In short, an argument is a finite sequence of truth-value susceptible expressions out of an object-language,[2] open to a variety of metalinguistic predicates, severally and sometimes jointly, such as "deductively (in)valid", "inductively (in)valid", "dialectically (in)valid", etc. This approach, though it is by no means the only way to proceed within the requirements we suggest, at least shows how we can be spared the necessity of ever saying that a bad argument is no argument at all, or that a good argument is good from all points of view [233].

Arguments are composed of statements, and in so saying we have a useful economy that allows us to inject some substance into a thesis by Hamblin; "A fallacy is a fallacious argument". We might indeed have some serious reservations about losing the fallaciousness of "Have you stopped beating your wife?" and the like, and perhaps should be ready to modify the ruling in the face of well-defined, significant exceptions. But the requirement is by no means arbitrary or capricious, and the justification of purported exceptions carries a heavy burden of proof.

What are we to say, then, of the typical sort of example advanced in introductory logic texts as an instance of the *ad baculum*,

(1) Shut your face or I'll kick it in?

Its untowardnesses are multiple - in varying contexts it might be illegal, immoral, or undiplomatic, and it is certainly impolite - but what *logical* sin does it commit? A succinct and pointed answer is that from the logical point of view this is not an argument - much less a correct or incorrect argument - it is a threat. And there is a syntactical initiation of the point: one of its constituent sentences is neither true nor false, not a declarative sentence at all, but an imperative [216, p. 5]. Much the same can be said for many a typical text-book examples of the *ad baculum*.

2. PRUDENTIAL INFERENCE

Now it might be counterclaimed that the *ad baculum* does indeed exhibit some sort of logical structure, even though its framework is not alethic. The suggestion may be that the setting of this fallacy is that of imperative inference, prudential argument, the so-called practical syllogism, or perhaps some other non-alethic context such as deontic logic or decision theory. In fact, frequently the text-book examples of *ad baculum* display a kind of deontic-prudential logical form. Consider two sequences presented by Alex Michalos as examples of the *ad baculum*.

(2) Either I'm right or you don't take the car tonight. Therefore I'm right.

(3) If it's your move I'll quit. Therefore, it's my move.

Let B be the bad state of affairs, some state clearly undesirable for the person to whom the sentence is directed. In (2), B = 'You don't take the car tonight,' and in (3), B = 'I'll quit.' Let 'I'm right' be p, and assume that p ranges over the values 1 and 0, respectively, the *should-be-brought-about* and the *shouldn't-be-brought-about*. Assume further that a value-assignment, V, meets the following condition,

(4) $Bp \supset V(p)=0$,

namely that the bad state should not be brought about. Then the form of (2) may be represented,

(A2) $p \vee B$
       Therefore, p

We can see that the form (A2) is somewhat akin to the classical disjunctive syllogism, since we are assuming that [~B] obtains. So (A2) could also be represented as,

(a 2*) $p \vee B$
        $\neg B$
        Therefore, p

We also assume as in truth-functional alethic logic,

(5) $V(p \vee q)=1$, where at least one of [p, q] has the value 1.

and

(6) $V(\neg p)=1$, where p does not have the value 1.

Accordingly (A2) is a valid argument form, and (2), its substitution instance, is likewise valid, though neither the premiss nor the conclusion is a statement.

The same pattern of reconstruction captures the structure of (3). Let p be 'It's my move', let q be 'It's your move', and supply the enthymeme, 'If it's not your move, it's my move'. The resultant form is

(A3) $q \supset B$
       $\neg q \supset p$
       Therefore, p

An appropriate semantic requirement on ' $\supset$ ' assures us of the validity of *modus ponens* and *modus tollens*, and vindicates both (A3) and (3). It can thus be shown that (2) and (3) do constitute arguments or something very like arguments.

This much we are prepared to concede. But what is not clear is how the reconstructions (A2) and (A3) reflect the fallacy of *ad baculum*, since, amusing to note, (A2) and (A3) are valid, not fallacious. Ironically, paradigm examples of the fallacy have come out of what seemed an appropriate fragment of semantical machinery as valid arguments. What then has happened to the fallacy?

Our persistent seeker after the *ad baculum* might respond as follows: There is all the difference in the world between correctly inferring or arguing that something is imprudent, and uttering a threat. The *fallacy lies in the specious use of the threat of force to escalate the belief of the person to whom the argument is directed that the conclusion is imprudent*. For example, there is a crucial difference between the two occasions of utterance of the sentence,

(7) Masami Tsuruoka is thinking of hitting you with his *shinai*.

according to whether the utterer is, or is not, Tsuruoka. In the former case, we may well have a threat, whereas the latter is more likely to be a warning, simply a non-fallacious notification that B is about to occur.

Well and good, then. There is a difference between a threat and a warning.

3. AD BACULUM AS EQUIVOCATION

Yet another suggestion for salvaging the *ad baculum* might be found in the confusion of two distinct questions, (i) the truth of a conclusion, and (ii) the prudential question of what might happen if the person to whom the "argument" is directed does not act in a certain way. Thus, of his two cases here met with earlier, Michalos says:

> In each of these cases, if the premises prove anything, it is that some kind of force is going to be applied unless a certain view is accepted. Or, to put it in a slightly different way, a certain view is going to be accepted or else. But the threat of force or violence is beside the point. The question at issue, say, in the last example is not what happens if it's your move, but whose move it is. The appeal to force is, from a logical point of view, an irrelevant (though often persuasive) appeal [137, p. 369].

And so it is suggested that the fallacy consists in confusing (i) whose move it is, with (ii) what happens if it's your move. What this may amount to is the thesis that the fallacy resides in the spurious conflation of the deontic-prudential (A3) with its classical truth-functional counterpart, (B3).

(A3) $q \supset B$
      $\neg q \supset p$
      Therefore, p

(B3) $q \supset \neg r$
      $\neg r$
      $\neg q \supset p$
      Therefore, p

The analogy is made more perfect by recalling that through (4), (A3) assumes that $\neg B$. Here we have two look-alikes, even to the extent that the conclusions appear to be identical. But they are not. The one on the left asserts that p is *true*; the one on the right asserts that p *should-be-brought-about*. What has happened here is a classical case of equivocation. Syntactically the arguments are similar, but semantically they are quite different. But where is the fallacy in this, if both (A3) and (B3) are, in their respective semantics, *valid*? Well the plain answer, we take it, is that (A3) and (B3) can be fallaciously confused. That is, when the fallacy-perpetrator advances (3) he intends that his victim will accept the conclusion of (B) on the evidence offered in the premises of (A3). This would indeed be a fallacy, an egregious one at that, but does it deserve the name of "*ad baculum*"?

We think not. This fallacy, as virtually all writers agree, consists in the *threat of force* or *violence*. And if I argue that p is imprudent, in a way designed to convince you that p is false, how am I guilty of the threat of force or violence? Of course, my imputation of imprudence might have had overtones of menace, violence or force, and perhaps then I would have committed the classical *ad baculum*, but the juxtaposition of (A3) and (B3) is simply neutral with regard to this additional factor. We conclude that *some* fallacy has here been captured, but that the *ad baculum* is still at large.

4. THREATS AND BELIEFS

A threat is not usefully thought of as valid or invalid, or true or false. A threat is effective or ineffective. What it means to say that a threat is effective is, roughly, that it moves the person to whom it is directed to or away from some specific course of action. The threat of violence is generally a powerful emotional factor, and it may play an important role in the modification of belief. *Ad baculum*, the appeal to fear, as with *ad misericordiam*, the appeal to pity, is first and foremost an instrument for the passions.

Aristotle, in the *De Sophisticis Elenchis* (174a 16) [8] pointed out that one very effective resource in refutation of an opponent is the strategy of stimulating his anger and promoting him to contentiousness. Aristotle writes that when agitated everybody is less able to take care of himself. Elementary rules for producing anger are to "make a show of the wish to play foul and to be altogether shameless". And, of course, unloosed emotions can in their own right be a strikingly effective positive factor in persuasion. Such a device was one of Hitler's favourites. "As a speaker, Hitler never tried to prove his assertions; he used statements as triggers for emotion ... He thought of an audience as a woman who was to be first emotionally aroused and then seduced and made to yield. "The last eight to ten minutes of a speech", said Ernst Hanfstaengl, "resembled an orgasm of words" [82, p. 276], [246].

The effectiveness of obvious blunt and brutal appeals to force in argumentation needs little defence, but it is not often recognized that very subtle forms of *ad baculum* appeal can also be effective (in a committee meeting or panel discussion, for example). There is the psychological phenomenon of the bandwagon effect. Certain members of any committee are likely to have varying degrees of subtle control over other members, and the value of such controls in influencing argument is perhaps greater than commonly realized [175].

We are getting now to the heart of *ad baculum*, but from the logical point of view it is a heart of darkness. True, the use of the appeal to fear is an important aspect of the psychology of belief-modification, of propaganda and the influence of opinion [135].

But the study of emotional factors in the modification of beliefs is the province of psychology, not of logic. True, while there is indeed a very necessary psychological aspect of actual, ratiocinative argumentation, the theory of the informal fallacies, as usually and optimally conceived, should not be reducible simply to psychology or rhetoric.

Copi writes that the fallacy of *ad baculum* is committed when force is used to cause a person to accept a conclusion [40, p. 74]. But where is the argument here? If I point a revolver at your temple to win your acceptance of my view, what premiss do I advance? Again, I might use "force" to cause you to accept or believe a conclusion, if I subject you to brainwashing or brain surgery, yet not advance an argument or commit a fallacy, or breach of argument, at all. Let us say at once that this offensive psychologism[3] is not a unique deficiency of Copi's treatment. It is altogether the standard treatment of *ad baculum* in the texts.

5. CONCLUSIONS

So far, we do not appear to have found any genuine instances of the *ad baculum*. To meet the requirements for an instance of *ad baculum*, a sequence would have to be (i) an argument, (ii) a fallacious or incorrect argument, and (iii) a threat or appeal to force. It may be possible to produce something that would meet all three requirements, but we will not scruple to say that it would take the rationality of a Tertullian commingled with the black power of a Svengali actually to pull the thing off - actually to commit the fallacy of *ad baculum*. Needless to say, ours is a conjecture quite at odds with the usual treatment of *ad baculum* in the texts, by which the fallacy is no *rara avis*. But until the thing is produced, we remain unconvinced.

Chapter 5
Ad Hominem

The informal fallacies have been a staple of the texts, manuals and curricula of centres of higher learning since Aristotle, but the prevailing interest in the fallacies has been rather more anecdotal and taxonomic than theoretical. The sad history of the subject is chronicled in Hamblin [80, ch. 1]. The worst of it is that what Hamblin disapprovingly calls The Standard Treatment is still the norm. In the teaching of logic, therefore, one seeks in vain for clear and adequate theoretical models for the fallacies, and this, of course, inhibits their use by students as an effective, to say nothing of decisive, tool in the analysis of argumentation [230]. To be sure, the informal fallacies are *informal*; and the tautology certainly prompts one to think - see [233], [234], and [224] - that the fallacies often outrun standard logical formalisms. But an informal fallacy is traditionally, and justifiably we think, an invalid argument, and where the invalidity of argument is concerned, some clearly articulated decision mechanism, some theoretically adequate model, should be looked for. As we see it, the basic stumbling block is the obvious lack of theory. The *ad hominem* is clearly one of the most important major informal fallacies, and we undertake here to set out a basis for an understanding of its underlying logical structure. After certain preliminaries we shall distinguish two main categories of *ad hominem*, the *tu quoque* and the abusive, and shall argue for the view that there are four categories of *tu quoque* of special significance.

1. THE LOCKEAN VIEW

The Oxford Dictionary attributes to Locke the first use in English of the expression of *ad hominem*. Where the phrase originated is not known, although Hamblin [80] conjectures that it originates from Aristotle [10, 177b 33, 178b 18, and 183a 21]. Hamblin quotes an interesting passage from the *Essay* [125] where Locke describes the *ad hominem* as a mode for refutation or persuasion by "pressing a man with the consequences drawn from his own principles or concessions". [125, p. 278f.] Locke contrasts the *ad hominem* with "the using of proofs drawn from any of the foundations of knowledge or probability", which he calls *argumenta ad judicium*. The concept of *ad hominem* formulated by Locke was consciously used by Galileo and referred to by the phrase *ad hominem* at least three times in the latter's major works. As Finocchiaro [63] points out, the notion was available to Locke from an explicit source much closer to Locke's time than Aristotle. Wherever it originated, here then is a distinction between modes of argument based on the kind of premiss chosen, though, to be sure, nothing in this account requires the distinction to be disjoint. The methods of deduction appropriate to each argument-type would presumably be the same. Hamblin rightly suggests that the Lockean distinction is rather similar to the more familiar contrast than Aristotle made between "dialectical" and "demonstrative" arguments.[1]

The first clue that the Lockean account does not accord with the idea of the *ad hominem* normally conveyed in modern texts is that, as Locke describes it, an *ad hominem* argument is not necessarily a fallacious argument. It is often appropriate, reasonable, and logically unexceptionable to press a man with undesirable or unforeseen logical consequences of his assertions.[2] It is sometimes counter-claimed that "contentious" argument is *always* sophistical. Perhaps the etiquette of argument is breached by one who avoids an issue by over-persistence with his opponent's premisses. Still more serious misdemeanours might be involved. But that *any* instance of such behaviour should be deemed a fallacy *ad hominem* we do not accept. Here is a view that virtually forecloses upon the possibility of the critical analysis of hypotheses. A Socratic argument is not by nature fallacious; and no one would seriously wish to deny Galileo's gift for teasing out interesting and often devastating consequences of the views of others in ways that exemplify a legitimate and constructive use of techniques of logical deduction. One of the most famous examples is his critique of the Aristotelian view that the velocity of a falling body is proportional to its size.[3]

SALVIATI: If then we take two bodies whose natural speeds are different, it is clear that on uniting the two, the more rapid one will be partly retarded by the slower, and the slower will be somewhat hastened by the swifter. Do you not agree with me in this opinion?

SIMPLICO: You are unquestionably right.

SALVIATI: But if this is true, and if a large stone moves with a speed of say, eight while a smaller moves with a speed of four, then when they are united, the system will move with a speed less than eight; but the two stones when tied together make a stone larger than that which before moved with a speed of eight. Hence the heavier body moves with less speed than the lighter; an effect which is contrary to your supposition.

Perhaps little more needs to be said to show that the *ad hominem* under Locke's description of it can be a correct and impeccable method of argument.

But in that view can the *ad hominem* be a fallacy at all? Certainly to argue from an adversary's premiss is a procedure that might be abused, or indulged in excessively, or confused with an appeal to independent evidence. But it is difficult to see how such abuses, even assuming them to be reasonably specifiable, should be identified with what is represented as the *ad hominem* in modern logic texts. Much the same disparity is suggested in the account of a Twentieth Century Lockean, Henry W. Johnstone, Jr., who goes so far as to propose in [110] that *all* correct philosophical arguments are *ad hominem*. According to Johnstone, philosophical arguments are typically indifferent or impervious to all evidence external to themselves, and thus that only consequences drawn internally from a philosophical view suffice for its refutation. Evidently Johnstone's view is an extension of the Lockean notion; each construes the *ad hominem* more broadly than modern texts allow.

2. THE STANDARD VIEW

We shall take as our touchstone of analysis the more standard contemporary view, namely, that the *ad hominem* is a fallacy of personal abuse or specious circumstantial refutation. We will eschew the Locke-Johnstone viewpoint since, as will presently emerge, it is too inclusive to capture what we consider the critical cases. Our own orientation reflects a determination to track down the *fallacy*. It is as a significant logical error, a snare and a delusion, that the *ad hominem* is interesting, a device whereby the measure, to use Bentham's terms, is defeated by impugning the man. Entirely typical forms of this type of argument are illustrated by Arnauld [11]:

> He is a man who owns nothing.
> Therefore, he is wrong.
>
> He is obnoxious.
> Therefore, he is wrong.
>
> He is quick-tempered.
> Therefore, what he says is false.

Writes Arnauld, "(t)he two aspects of discourse, matter and manner, must be considered separately. The manner must be judged by the manner; the matter, by the matter; and not the matter by the manner nor the manner by the matter" [11, p. 291]. It is to the modern conception of the *ad hominem* typified by these examples from Arnauld that we shall direct our attention in this essay.

3. THE CIRCUMSTANTIAL AD HOMINEM

The circumstantial *ad hominem* or *tu quoque* is usually said to be committed when an inconsistency exists between a statement and the circumstances of its utterance. There appears to be a certain latitude regarding what counts in regard to the *tu quoque*, as "circumstances", but the tenor of most contemporary accounts suggests that what is common to these factors is a *personal orientation* pertaining to the behaviour of the individual who happens to be the author or source of the statement. Typically the charge of *tu quoque* is levelled at the individual whose exhortations are belied by his own behaviour. One timely example is this: a Citizens' Committee protests the Government's proposal substantially to raise the salaries of Members of Parliament on the ground that the Government has recently entreated the Canadian public to resist such inflationary practices as wage-escalation. Thus Archbishop Whately [201]:

> ... in the "argumentum ad hominem" the conclusion which is actually established, is not the absolute and general one in question, but relative and particular; viz. not that "such and such is the fact", but that "this man is bound to admit it, in conformity to his principles of Reasoning, or in consistency with his own conduct, situation", etc. (p. 196).

De Morgan, in [50], makes much the same point: that the fallacy of *ad hominem* generally has some reference to the particular person to whom the argument was addressed. Many modern accounts, including Copi [40] and others too numerous to mention, share in this tradition.

Philosophers such as Rescher [159] see a significant distinction between the *tu quoque* and the circumstantial *ad hominem*. We ourselves shall, however, seek to specify four species of this general and unbifurcated type of argument, each a case of the *tu quoque* or the circumstantial *ad hominem*, as you please.

4. LOGICAL INCONSISTENCY

Often, as De Morgan suggests, the argument against the man consists in recrimination and a charge of inconsistency, as, 'You cannot use this assertion, because in such another case you oppose it.' Taking up De Morgan's idea, one medium of the *ad hominem* is *assertion*, where we understand assertion to have at least the following characteristics. (a) An assertion is characterized as a relation between an individual, a, the *assertor* or *source*, and a statement, p. (b) Assertions, as we conceive them, are localized as to time and place of utterance. (c) Like belief, and unlike knowledge, assertion is not truth-entailing. It is open to one to assert falsehoods and inconsistencies. (d) Assertion is not closed under consequence. If I assert p, I do not necessarily assert all the deductive consequences of p. Normally, however, I will be held to have asserted *some* deductive consequences of p in asserting p; but just which consequences these are we will not here try to specify. (e) Assertion is one of those modalities concerned with a broadly psychological attitude concerning an individual and a statement. The question whether a has or has not asserted p is, to some degree, a behavioural question. Other modalities in this category which are also relevant to the analysis of the *ad hominem* are *statement, proposition, acceptance, belief*, and *inference*.[4] Each of these yields a variant of the first account of the *tu quoque* below, when substituted for the assertion modality:

Consider now the case in which: (TQ) a asserts that p & a asserts that $\ulcorner \neg p \urcorner$. Is this the fallacy whose analysis we seek? True, inconsistency is the deadliest logical sin, but some inconsistencies are more serious than others. A lifetime of total consistency could be as much a symptom of imbecility as logical acumen (to say nothing of the crushing boredom). Some inconsistencies are rather easily removed by the sacrifice of a relatively inconsequential part of the assertion, or by the addition of minor qualifications. True, as Perelman and Olbrechts-Tyteca [143, p. 206] point out, a person who is seen to sin against logic is ridiculous at the outset, "assuming that he is not considered insane or so lacking in credibility that nothing he did could disqualify him more". A person who can be accused of maintaining two demonstrably incompatible points of view is uniquely vulnerable to ridicule.[5] In making an argument, one advances considerations in behalf of a position, and does so in such a way that, if one is successful, it is unreasonable for him, to whom the argument is directed, to resist the argument's conclusion. The arguer, therefore, displays an overriding commitment to reasonableness. It is as if he appeals to his interlocutor, "Don't you see? If you do not accept what I say, you're inconsistent". But if his *own* argument commits the deadliest logical sin, the interlocutor may say (and may say it with justifiable relish), "Inconsistent? I? Look who's talking". *Tu quoque.* Obviously, *this* is no fallacy.

Imagine, however, that the inconsistency embodied in (TQ1) is easily removed, without damage to the position to which the argument is directed. Then, to hold a man to his inconsistency, to ignore its effortless eliminability, is to win your point on the merest technicality. It is to captiously subvert the principal objectives of *reasonable* argument to the ambitions of successful advocacy. It involves a cavil which deserves to be called "fallacy", but is the fallacy the *ad hominem*? Yes, for it manifests a very common attribute of the *ad hominem*. The charge is accurate, but is also irrelevant or frivolous. (We shall say more of this in § 11).

The *reductio ad absurdum* can also be an *ad hominem*, where a asserts a well-evidenced proposition p conjointly with a relatively trivial adjunct q, and his opponent, b, derives $\ulcorner r\ \&\ \neg r \urcorner$ from $\ulcorner p\ \&\ q \urcorner$, spuriously inflating the seriousness of the contradiction even though the retraction of q would root out the inconsistency. The fallacy here committed reminds one of the analysis of the classical Aristotelian Non-cause as Cause[6] proposed by Hamblin [80], which consists in confusion between the correct form:

(1) $((p \rightarrow q)\ \&\ \neg q) \rightarrow \neg p$

and its incorrect look-alike:

(2) $((p\ \&\ q) \rightarrow r)\ \&\ \neg r) \rightarrow \neg p$

This is evocative of the Stoic fallacy of superfluous premisses - a superfluous premiss can be tacked on to an otherwise correct argument to yield an inconsistency. Significantly, however, the question of whether a premiss is superfluous or relatively trivial is, of course, not one that can be decided in classical first-order logic. A valid argument remains valid no matter how many additional premisses are "tacked on", and similarly, as (2) indicates, the derivation of an inconsistency from a set does not locate the specific source of inconsistency within that set. Of course further exploration of the fallacy relative to (TQ1) may find that relevant logics offer a congenial home - see Anderson and Belnap [2], Lehrer [121], and Woods and Walton [235]. Lehrer connects the notion of superfluity of premisses with the notion of relevance in an interesting and possibly very fruitful way for current purposes. But we do not pursue these suggestions here.

It is clear that a complete understanding of the kind of fallacy of *ad hominem* associated with (TQ1) can be achieved only in a model that allows the notion of repudiation or retraction of one's previous assertion or commitment. Essentially cumulative models of argumentation, such as the Obligation Game[7] of Hamblin [80, 260-63] or the Kripke semantics for intuitionistic logic [117], do not therefore fully reveal the contours of the *ad hominem*. A simple illustration would be the 'Why-Because System with Questions' of Hamblin [80, p. 265ff.]. This is a simple dialectical system (game of logic, dialogical game) in which two participants take turns in making certain moves specified by a set of rules. However, unlike the Obligation Game, the Why-Because Game allows for deletions of propositions from the speaker's commitment-store. Thus the following sort of sequence might occur in a specimen of dialogue. Woods asserts A, but then Walton, who is good at these things, demonstrates through a series of steps sanctioned by the rules (in this case a set of rules for the propositional calculus) that $\ulcorner \neg A \urcorner$ is a consequence of B, a statement in the commitment-store of Woods. Walton then infers the falsehood of Woods' initial thesis, A. Or, in a weaker move, Walton might conclude that Woods was incorrect in asserting A. Now we would like to suggest that in either case, Walton could fairly be accused of having committed a form of the *ad hominem* fallacy against Woods. Our reasoning is as follows. Walton has fairly, let us say, caught Woods in contradiction - he showed that B, which Woods is committed to, entails (by rules which both accept) the negation of another proposition that Woods also accepts. So far Walton is unimpeachable, unlike Woods who is caught in inconsistency. But Walton is unreasonable to conclude that Woods must give up A. On the contrary, Woods might well decide to retract his commitment to B, thus avoiding inconsistency. Nor is Walton justified in the even stronger conclusion that A is false, since of course, simply because $\ulcorner \neg A \urcorner$ follows from a commitment of Woods, it does not follow that A is false or unjustifiable, apart from Woods' commitment. The point is that in the Why-Because Game, Woods can retract his commitment to B, and consequently can continue consistently to maintain A, if he opts to make those moves. Walton argues fallaciously by attempting to block that option prematurely.

Walton argues fairly in pointing out the inconsistency, but unfairly in attempting to prevent Woods from removing the inconsistency by retraction of a commitment, a move that should be open to him in the sort of game we describe.

Of course, in the monolectical framework of first-order logic, any proposition you like follows from an inconsistency. And indeed, any essentially cumulative or exclusively incremental model of argumentation will not fully reflect the fallaciousness of (TQ1) *ad hominem*. It is only in the dialectical setting of a two person logic with provision for the retraction of commitments that the fallacy can be adequately understood as a species of incorrect argument.

5. ASSERTIONAL INCONSISTENCY

We mentioned above, and we will elaborate the theme further below, that the inconsistency cited in a charge of *ad hominem* is typically that between an assertion and some action or circumstance of the assertor. This emphasis suggests that some performative inconsistencies such as those studied by Hintikka [95] and Walton [198], and self-refutations as studied by Mackie [131], may be instances of arguments *ad hominem*. This suggests the general schema:

(TQ2) For any logically consistent p, for any a, for any time t, p is *assertionally inconsistent* where it is logically inconsistent that (1) a asserts that p at t, and (2) p is true.

For example, the statement 'I now assert nothing' is assertionally inconsistent for me to assert now.[9] As before, there is no fallacy in such allegations unless the inconsistency is open to cancellation without damage to the point at issue. Only a fool will seriously jib at "Oh, I never talk", unless it is clear that there is no place for the usual interpretations of such disclaimers, interpretations such as "Oh, I'm not very talkative", or "I seldom take much of a role in conversation". Many varieties of (TQ2) are possible and interesting to posit by substituting in (TQ2) other relevant modalities for *assertion*, e.g., *statement, proposition, acceptance, belief, inference*. Thus we might wish to consider inferential inconsistency, doxastic inconsistency, and so forth. These varieties of inconsistency provide a medium for the *ad hominem* that link the previous cases of purely logical inconsistency with another kind of inconsistency, germane to the *tu quoque*, to which we now turn.

6. PRAXIOLOGICAL INCONSISTENCY

A third species of inconsistency is of a praxiological or action-theoretic sort:

> (TQ3) For any logically contingent p and any individual a, p is *praxiologically inconsistent* where a asserts that p and a brings about that ¬p.

This kind of inconsistency easily reduces to logical inconsistency provided we accept the plausible axiom, (ad) if a brings it about that p, then p. For if a brings it about that $\ulcorner \neg p \urcorner$, and by (Ad) $\ulcorner \neg p \urcorner$ thereby obtains, then a has in effect "asserted" that $\ulcorner \neg p \urcorner$ through his actions (do actions speak louder than words?). But if a simultaneously asserts p verbally, he contradicts himself. Thus (TQ3) characterizes the case where what a man says is inconsistent with what he does, provided, of course, that the man "knoweth what he doth".[10]

7. DEONTIC-PRAXIOLOGICAL INCONSISTENCY

Most often, however, allegations of the *tu quoque* turn on an alleged inconsistency between what a man does and what he says *should* be done. To avoid this recrimination, a man must "practice what he preaches" (and conversely).

> (TQ4) For any logically contingent p and any individual a, p is *deonto-praxiologically inconsistent* where a asserts that p should not be brought about and a brings it about that p.

A stronger variant of (TQ4) is achieved by altering the *definiens* to read: a asserts that $\ulcorner \neg p \urcorner$ should be brought about and a brings it about that p.

A particularly interesting instance of *tu quoque* of the (TQ4) sort is the notorious *hunters argument* due, it appears to Archbishop Whately [208]. Consider the sportsman accused of barbarity in his sacrifice of hares or trout for his amusement. 'Why do you feed on the flesh of animals?' he replies. By Whately's lights, this is an *ad hominem* argument and a *correct* one. Whately also noticed that this *ad hominem* often has the effect, not unreasonably, of shifting the burden of proof to an unfairly aggressive adversary. Now connecting the *ad hominem* with the notion of *burden of proof* is an excellent and fruitful suggestion, and is usefully pursued by Hamblin [80]. But the question arises: if the *ad hominem* is sometimes a correct form of argument, then what differentiates the correct *tu quoque* from the fallacious? Whately replied: "the fallaciousness depends on the deceit or attempt to deceive" [208, p. 197]. For our part, we cannot accept this answer. Whether an argument is correct or fallacious surely must not be held to be question of the honourable or deceitful intentions of the arguer. Here lies the route to an altogether unwelcome psychologism. A fallacy is an *incorrect* argument, not merely a correct argument used unscrupulously.

De Morgan's response to Whately is more than historically interesting and indeed worthy of close scrutiny. Disdaining Whately's psychologism, De Morgan neatly explained the fallacy in the hunters example as a illicit substitution into the following alleged inconsistency: "The parallel will not exist until, for the person who eats meat, we substitute one who turns butcher for amusement" [50, p. 265]. In our terms, what has happened is that the p occurring in one side of the inconsistency-generating conjunction in the *definiens* of (TQ1), (TQ2), (TQ3), or (TQ4), is not the same expression that appears as p in the remaining conjunct. Take the *definiens* of (TQ4) as an example:

> *a* asserts that hunting should not be brought about and *a* brings it about that meat-eating obtains.

There would be no inconsistency, even of the deontic-praxiological variety, in condemning butchery for amusement over a steak dinner. Hence the fallacy here can be identified as an illicit substitution into (TQ4), or, more simply, as the ensuing ambiguity: "... it is not absolutely the same argument which is turned against the proposer, but one which is asserted to be like it, or parallel to it. But *parallel cases* are dangerous things, liable to be parallel in immaterial points, and divergent in material ones". [50, p. 265]. Modern text-book writers would have done well to consult De Morgan. None that we are aware of ever did. Copi [40] offers a standard non-explanation of the hunters argument: "Arguments such as these are not really to the point; they do not present good grounds, etc". One is tempted to argue, *tu quoque*, that this explanation is not really to the point, or even near it.

We now turn to the second main category, the abusive *ad hominem*.

8. THE ABUSIVE AD HOMINEM

An initial problem is that abuse does not even constitute an *argument*, let alone a fallacious argument.[11] However, arguments can be abusive, and where the abuse constitutes an *ad hominem*, it usually calls into question someone's credibility, by alleging deviances or shortcomings in the person's authoritativeness or expertise. Thus the abusive *ad hominem* is the inverse of the *ad verecundiam*; and we will here sketch the notion of expertise as a preliminary for what follows.

9. EXPERTISE

Wherein lies the expertise of an expert? A rough answer can be framed as follows. Someone is an expert in a domain, d, only if there are good reasons to believe that his predictions in d will have a significantly greater proportion of successes than that of a layman. Now 100% accuracy in predicting the rising of the sun does not one a meteorologist make. Simple accuracy is not enough. An accuracy significantly beyond that of a lay predictor is required. But how significantly?[12] It would be naive to expect an *a priori* answer to this, the question of the measurement of degree of expertise, although we might expect such measures eventually to emerge from empirical studies on the technology of expertise. We might also expect these measures pragmatically to vary with such parameters as the domain, d. Still, as we suggest in [225], two factors will be prominently relevant to the measurement of degrees of expertise - a domain, d, and a layman-relative success factor, say, k.

Even aside from questions of measurement technology, the foregoing sketch suggests two basic conceptual problems. First, we normally associate expertise with underlying theoretical competence rather than simply with an edge over the layman in actual performance. The simple edge over a layman might be due to luck or some other factor not essentially related to expertise. A test case is provided by the hypothetical carnival fortuneteller, Madama Fortuna, who, for all her midway shenanigans, astonishes the world, and herself, with an extraordinary record of success in forecasting the weather. It would be unwise to ignore Madama Fortuna's prediction when planning a picnic, but would we want to allow that she is a true meteorological expert? If not, we concede the existence of some other factor in expertise beyond a statistical edge on the lay predictor. The tension exhibited here between purely statistical factors and other factors evidently relating to underlying theoretical competence, affects the structure of the logic of expertise at other points as well. For example, it requires further niceties in the concept of expertise to square the nonsentential aspect of an expert's inarticulable store of background knowledge (in, say, cases of expert medical diagnosis) with the perfectly reasonable demand that some sort of objective evidence be forthcoming from an expert who diverges in opinion from other experts.[13] It thus appears that we require of a *bona fide* expert not only (i) a statistical success factor, but (ii) a factor of underlying knowledge.

We have argued in [225] that there do exist rational criteria for the adjudication of appeals to expertise, and, unquestionably, some appeals to expertise are instances of legitimate forms of argument. Nevertheless, where direct access to data becomes available, the appeal to expertise may loose its usefulness, and even its relevance. Direct appeal to evidence is, to be sure, always to be preferred. On the other hand, with the extraordinary growth of specialized knowledge, it is increasingly necessary to recognize a proper place for the appeal to authority.

10. NEXPERTISE

Salmon [167] has pointed out that certain kinds of *ad hominem*, namely those he calls "argument against the man", reduce to a special case of the statistical syllogism, namely one in which appeal is made to *lack* of expertise:

(P1)   The vast majority of statements made by x concerning subject S are false.

(P2)   p is a statement made by x concerning subject S.

(C)   Therefore, p is false [167, p. 68].

We will call an individual who fits the above scheme a *nexpert*[14] a predictor who is even more likely to be wrong than a layman. Thus, in general, it would seem that the class of predictors admits of a tertiary partitioning, as follows:

Expert	Layman	Nexpert
	Non-expert	

Both lateral categories are defined relative to the layman who, as we might expect, would normally achieve a somewhat better than 50% accuracy in the usual case. We expect the expert to do better, and the nexpert to do worse. Thus we might have a weaker and stronger variety of *ad hominem* - we might weakly impugn a man's credibility by calling him a non-expert, or more strongly, we might accuse him of positively being a nexpert. Thus a non-expert is either a layman or a nexpert.

Salmon's point is helpful, and yields a new perspective on the *ad hominem*, but we think his insight could be deepened by altering the conception of (P1). One's grounds for believing reasonably that the pronouncements of a nexpert will likely be less accurate than even that of a layman need not reside exclusively in considerations of track record, as is suggested by (P1). Generally our distrust of the nexpert is no simple matter of his having been wrong in the past (although, admittedly, this is an important factor) but more because of his theoretical peculiarities and shortcomings. The nexpert is working with a deviant or deficient theoretical apparatus. Like its obverse, expertise, nexpertise ranges between the two poles of actual track record and underlying theoretical competence, and Salmon's account is too weighted at one end.

As might be expected, nexpertise manifests itself in a variety of poses, each contributing in slightly different ways to deficiency of argument. One kind of nexpertise is *topic selective*, that is, it involves a relevantly low degree of information and understanding about the particular topic to which the argument in question pertains. In its crispest form, topic selective nexpertise makes for a corruption of argument when: (i) the argument deals with some topic or domain of enquiry d; (ii) competent argument with respect to d requires the possession of some specialized knowledge or experience, rather than what we associate with general intelligence and standard well-informedness; and yet (iii) our nexpert lacks this special expertise. To any substantive argument he might make about matters in d, there is a perfectly correct rejoinder *ad hominem*: "But you don't know anything to speak of about d". Typically, such a rejection is agnostic in force.

A(1) X asserts that p, q, r about *d* and concludes that s.

A(2) But, though no expert about *d* myself, neither is X.

A(3) And compelling arguments pertaining to *d* require special expertise.

A(4) Therefore, we are not justified in accepting X's conclusion, even though, for all we know, it is correctly inferred from correct premises.

Of course the (perfectly proper) rejection of an argument on grounds of topic selective nexpertise can also occasionally have a stronger than agnostic force - let us say *atheistic force*.

B(1) X asserts p, q, r about matters in *d*, and concludes that s.

B(2) But X is so considerably misinformed about *d* that it is practically certain that any of his substantive assertions about *d* will be wrong.

B(3) Consequently, it is justifiable to hold that this particular argument is also wrong.

It is worth remarking that the complaint embodied in argument A(1)-A(4) is that the argument about which it protests fails because (i) given its topic *d* the argument could succeed only if, to some extent at least, it incorporated the benefits of a *non-fallacious ad verecundiam*; and (ii) the arguer lacks the expertise for a non-fallacious *ad verecundiam* in domain *d*. We shall have occasion below to say something further about the inter-relationship between the *ad hominem* and the *ad verecundiam*. For now let us merely add that argument B(1)-B(3) is itself a good argument contra the argument of which it complains only if it is made by someone who himself is enough of an expert about domain *d* to be justified in his confidence that Mr. X, about whom he protests, is almost certain to be wrong in any of his pronouncements about *d*. So something of the justified *ad verecundiam* seems to be involved in the *success* of B(1)-B3), as opposed to the *failure* of the argument against which A(1)-A(4) is directed. Arguments of the B(1)-B(3) sort also risk an interesting kind of stand-off, often associated with clashes of dogma, as with a doctrine of "establishment" science and competing doctrine of "revolutionary" science. Here the particular point of disagreement pertains to associations of the form "Anything he says about d is bound to be wrong", wherewith the quite real possibility of a mutual begging-of-the-question - as one would rightly enough expect of dogmatic disagreements.

Another kind of nexpertise in *topic neutral*. It involves a quite general sort of incompetence, a general infacility either with argument or with making sense of the world. As before, *ad hominem* charges promoted by topic neutral nexpertise can quite soundly licence either agnostic or atheistic repudiations of the arguments to which they are directed, and can, though with less likelihood we would think, be the source of mutual beggings-of-the-question. The involvement of the *ad verecundiam* is also less clear. If you reject a man's argument because he is an utter simpleton about everything whatever, you neither complain that he is not an expert, nor need make the complaint on grounds of your own expertise. General incompetence is not so obscure a failing as to be recognizable only by experts.

Still another kind of nexpertise pertains to the epistemic control of one's premiss-store. It is one thing to get your assertion wrong about d because d-related matters are as such beyond your competence; it is another to get them wrong because of simple ignorance - because you have been misinformed, inattentive, lazy or careless. It is reasonable enough to counter-argue:

C(1)   X asserts that p, q, r about matters in d, and concludes that s.

C(2)   But, though d-related matters are not (or not likely) beyond X's capabilities, X nonetheless has got p or q or r just wrong.

That is, the complaint *ad hominem* can clearly sustain an agnostic rejection. However, an atheistic rejection would seem to lie open to a deficiency of its own, namely, the *ad ignorantiam*. For, granted, if X does not know what he claims to know, he is not entitled to claim to know that s is false.[15]

A somewhat different deficiency, one that scarcely justifies the appellation "nexpertise", has to do with an arguer's *character* or the *special circumstance of his advocacy*. If we are dealing with either a chronic liar, or, what is more likely, with someone who almost certainly would lie about matters in domain d, it is often justified to reject his argument - though with what degree of severity it may often not be clear. Similarly, if we are dealing with an opponent in an action before the courts or in a debate in or out of Parliament, then the special circumstances of advocacy make for a selectivity concerning evidence, and a general willingness to sacrifice truth to persuasion, which entitle me to appropriate kinds of hesitation about the inferential and epistemic success of such arguments. As before, various degrees of rejection may be more or less appropriate, with much the same opportunities for overdoing it, in the direction of the *ad ignorantiam*. And here, too, elements of the *ad verecundiam* would seem to have various roles to play. Thus, if I complain of your argument on grounds that you are (merely) a lawyer, I may have it in mind that you are not *exercising* the appropriate expertise, whether or not I think that you possess it.

11. CRANKS AND CRACKPOTS

Perhaps the best understood (or anyhow most obvious) example of the nexpert is the scientific crank, the pseudo-expert who uses the prestige and authority of established science fallaciously to escalate the credibility of his eccentric claims. Somewhat after the fashion of Gardner [69], the crank may be distinguished from the legitimate if unorthodox scientist by his possession of various of the following set of characteristics.

1. Wholesale Rejection of Authority: the crank not only rejects the standard authorities in a given field, but also often characteristically rejects all established science altogether.
2. Incompetence: the crank displays a vitiating ignorance of the areas of science he rejects.
3. Lack of Communication: there is no communication with the rest of the practitioners of his field. If he publishes articles, it is in his own journal, if books, they are paid for by himself or his friends.
4. Opposition of scientists to his views he regards as the bigotry of scientific orthodoxy.
5. The crank hails himself as a genius, misunderstood by the conventional plodders, and typically associates himself with certain towering figures such as Plato, Newton, or Einstein. These major figures, however, he usually misunderstands and misinterprets.
6. The superiority of some other doctrine as an authority-alternative to established science is

usually held - often the doctrine is religious, broadly philosophical, or moral. Usually, established science is held to be based on a deep conceptual error.
7. Often the crank has a cult, or entourage of disciples, to whom he appeals with heavily emotional expression and exaggeration. The history of thought in his area is dramatically dichotomized into "friends" and "enemies".
8. There is a strongly expressed belief in the capability of the leader of raising the intelligence of the disciples in the direction of genius. Typically, little extended effort is required in this undertaking - it is more a matter of instant insight. Often these claims accompany odd theories of education, mystical insight and so forth.
9. An urgent social application is often strongly suggested - "Only my ideas can save the world".
10. Often theories are accompanied by extravagant claims about their practical benefits. Wondrous physical or psychological benefits are frequently a key part of the crank's programm.[16]

That the crank is a familiar figure in the history of science does not make his recognition always a straight-forward matter. There are many borderline cases, some of them extraordinarily difficult to assess. Some alleged cranks[17] did make legitimate, recognized contributions to their area before turning to the invention of "cloud-busters" and "orgone generators". Others, like Korzybski,[18] are held to hover on the borderline of legitimate scholarship. It hardly needs saying that whatever the final word on crankdom, the crank captures our interest here for his nexpertise.

12. SOME FURTHER INTERCONNECTIONS

Suppose that a popular sports hero commits a fallacious *ad verecundiam* in behalf of the electronic marvels of a line of sophisticated computer equipment that he is advertising. Then in rejecting his argument for this reason, I commit a *non*-fallacious *ad hominem*. The general rule would seem to be this:

The correct rejection of an argument, for its having committed the fallacy of ad verecundiam, *involves the non-fallacious use of an* ad hominem.

Consider now that, in appealing to an utterly eminent intuitionist logician, I make use of a non-fallacious *ad verecundiam*. But you protest my appeal to authority and reject my argument. Your own *ad hominem* is incorrect. More generally,

> *The incorrect rejection of an argument, for its having committed the* ad verecundiam, *involves the incorrect use of an* ad hominem.

We said, just now, that there is involved an "incorrect" use of an *ad hominem* in order to have a sufficiently general term to cover two quite different deficiencies. One commits a *falsehood ad hominem* when the argument-impairing attribute is *falsely ascribed* to an arguer. One commits a *fallacy ad hominem* when one correctly ascribed the attribute, but *wrongly infers* that it wrecks the arguer's argument. Rather more strictly speaking, the *ad hominem* involves a fallacy (a logical blunder) as opposed to a mistake of fact, when the conclusion is incorrectly drawn that an arguer's attribute Φ wrecks his argument, A, irrespective of whether Φ is rightly ascribed. So conceived, what makes the *ad hominem* a fallacy has nothing essential to do with its "address of the man".

The attribution can be correct or incorrect, and whichever it is, any argument incorporating such attributions will be *ad hominem*. However, if the conclusion of the argument is not relevantly affected by one's possession of the putative attribute, then a fallacy has been committed, yet it is simply the fallacy of having made an invalid inference - it is a fallacy of relevance.

13. CONCLUDING REMARKS

We neglect the study of the fallacies at our peril, for it is just in these areas that rational criteria, however inexact and tentative, are sorely needed as an aid to the adjudication of actual, everyday argumentation. While the traditional treatments of the fallacies are too unsystematic to be useful as an effective device in argumentation, their abandonment leaves a gap that no one (yet) quite knows how to fill. Hamblin suggests that we are in the position of the medieval logicians before the 12th century. We have lost the doctrine of fallacy and need to rediscover it [80, p. 11]. Our current standards of theoretical rigour are somewhat higher than those of the 12th century, and therefore theories of fallacies are unwisely constructed entirely in isolation from other more systematic departments of contemporary logic. We are under no illusion that the treatment of the *ad hominem* here proposed is destined to become theoretically well-entrenched. It suffices we think, for all its tentativeness to help the dialogue to get started.

Chapter 6
Petitio and Relevant Many-Premissed Arguments

We want to examine the question whether it is possible for arguments with more than one premiss, and no superfluous premisses, ever to beg the question. In order properly to formulate this question, it is necessary (i) to set out a general account of the conditions under which an argument may be said to beg the question, (ii) to explicate the concept of a "superfluous" premiss, and (iii) to adumbrate a certain historical dispute which throws interesting light on the nature of the question.

1. PETITIO PRINCIPII

That arguing in a circle is an essential *informal* fallacy is easy to establish. Arguments of the form ⌜p, therefore p⌝ always or nearly always beg the question, yet their formal validity is impeccably reflected in standard truth-functional logic. Thus there is a kind of dissonance, especially palpable in the paradigm ⌜p, therefore p⌝, that pervades all arguments in which the conclusion appears as a premiss or conjunct of a premiss. Such arguments are both valid and fallacious.

We are recapitulate here from [234], our survey of what is standardly conceived to be the fallacy. In writings on *petitio* there are, broadly speaking, two main conceptions of this fallacy. According to the equivalence conception, an argument is circular where the conclusion is tacitly or explicitly assumed as one of the premisses, that is, where the conclusion is equivalent to a premiss. Copi [40], for example, writes [40, p. 83],

> If one assumes as a premiss for his argument the very conclusion he intends to prove, the fallacy committed is that of petitio principii, or begging the question. If the proposition to be established is formulated in exactly the same words both as premiss and conclusion, the mistake would be so glaring as to deceive no one. Often, however, two formulations can be

sufficiently different to obscure the fact that one and the same proposition occurs both as premiss and conclusion.

The problem with the equivalence conception is that strict identity of premiss and conclusion is too narrow a criterion, reflecting only the obvious cases, and equivalence is too broad a criterion, ascribing circularity to many arguments which clearly are not circular. De Morgan [50] preferred the narrower criterion that, "strictly speaking, there is no formal *petitio principii* except when the very proposition to be proved, and not a mere synonyme of it, is assumed" [50, p. 254]. Yet Sidgwick [173] was of the opinion that nothing "appears to be really gained by restricting the name to so small a compass as this; and there is no doubt that such a restriction would be very much at variance with the popular acceptation of the term" [173, p. 194]. On the other hand, the broader criterion, demanding only that a premiss be logicallly equivalent to the conclusion, would appear to condemn single-premissed arguments such as

All philosophers are fragile
Therefore, no philosophers are not fragile

to hopeless circularity in every context. Of course, it is sometimes said that an argument is circular when to state a premiss is to state the conclusion (in an appropriately epistemically rich sense of "state"). The trouble is, *does* one, in here stating the conclusion, in an appropriately rich sense of "state", also state the premiss?

According to what we call the dependency conception, an argument is circular when the conclusion is pre-supposed by a premiss or where some premiss actually depends on the conclusion, in the sense that in order to accept the premiss one need first accept the conclusion. Normally the "flow of inference" in an argument is from the premiss to the conclusion.

Figure 1

But where it is also required that an inference be made in the other direction, from the conclusion to the premiss,

Figure 2

the argument, schematized in Figure 2, is circular. According to this conception, an argument is non-circular only if one may know that each premiss is true without having to infer it from the conclusion, or from some other statement that can be known only by an inference from the conclusion.

Two additional general points bear remarking on. First, some writers are inclined to think that begging the question is a purely *alethic* question, pertaining to a relation of logical entailment (as dependency) or logical equivalence between the conclusion and a premiss. Others (notably Whately [208, p. 179]) explicitly recognize that circularity of argument is essentially an *epistemic* phenomenon, and that whether or not an argument is said to beg the question will depend critically on the informational or epistemic circumstances of the person to whom it is directed. Secondly, we have never seen in the logic texts an allegation of *petitio* where two or more premisses are said to jointly or severally beg the question. Whether cases of this sort are *possible* is of course a separate question that we return to below.

2. MILL: THE SYLLOGISM IS A PETITIO

Mill was concerned to differentiate between the mental operation of inference whose medium is belief, and the purely logical concepts of deductive proof and valid argument which are deemed timelessly to map truths onto truths in the doxastically neutral, characteristic way soon to be elevated to mathematical logic by Boole and Frege. The *System of Logic* is laden with Mill's attempts to grapple with this important distinction. Perhaps the most celebrated and provocative offshoot of Mill's preoccupation is his *dictum* that there is a *petitio principii* in every valid syllogism. The thrust of Mill's argument [138, p. 120f.], is easily felt. Consider the archetypal syllogism,

(1) All men are mortal.
    Plato is a man.
    ───────────────
    Plato is mortal.

Clearly the first premiss presupposes the conclusion in the sense that we cannot be assured that all men are mortal, "unless we are already certain of the mortality of every individual man". Thus if it is doubtful that Plato is mortal, or any other individual for that matter, then it is at least as doubtful whether all men are mortal. The point is that the general statement 'All men are mortal' cannot be offered without circularity as evidence of the particular statement 'Plato is mortal' when the latter is part of the evidence for the former.

Mill's *aporia* is no superficial puzzler. This can be seen from his ensuing discussion in which he observes that the typical reply of logicians involves an appeal to a distinction between what is *asserted* by the premisses and what is *implied* in the premisses. But is this distinction really all that clear? When Whately, for example, tries to deal with Mill's problem by claiming that when you admit the major premiss you assert the conclusion, but assert it by implication merely, does he mean that your asserted it unconsciously? This is a fascinating preview of contemporary disputes concerning rationality-assumptions in epistemic and doxastic modal logics, but we will not here pursue it further. We turn instead to the response of De Morgan.

3. DE MORGAN'S REJOINDER

De Morgan, that staunch defender of formalistic methods, was inclined to argue for what we will now call the Equivalency Reduction Thesis, namely, that all cases of *petitio* can be ultimately reduced to the case where the "very proposition to be proved, and not a mere synonyme of it, is assumed" [50, p. 254]. His preference for the equivalence conception is not merely aesthetic - we will see further below that De Morgan offers what amounts to an ingenious and deeply based defence of this apparently implausible position.

De Morgan opens his discussion of Mill's conundrum by pointedly ruling that the definition of *petitio* refers to what is assumed in one premiss [50, p. 257]: "The most fallacious *pair* of premisses, though expressly constructed to form a certain conclusion, without the least reference to their truth, would not be assuming the question, or *an* equivalent". This thesis of a connection between *petitio* and the number of premisses is not today commonly known or recognized, even in standard treatments of the fallacies, but it had been clearly stated before De Morgan [50]. Whately [208,p. 179] writes:

> *Petitio principii* ... takes place when one of the Premisses ... is either plainly equivalent to the conclusion, or depends on that for its own reception. I have said "one of the Premisses" because in all correct reasoning the two premisses taken together must imply and virtually assert the conclusion.

Yet because Whately only mentions the thesis in passing, and because De Morgan was apparently the first to attempt to establish it in a general way[1], we formulate the following principle

De Morgan's Thesis: No syllogism begs the question.

A syllogism is classically required to have exactly two premisses, and it follows that if it has exactly three distinct terms each of which occurs twice, one in both premisses and the remaining two in one premiss and in the conclusion, then neither premiss will be superfluous. Of course, the restriction of *De Morgan's Thesis* to specifically syllogistic arguments is a bit archaic, and one is tempted to formulate the thesis more generally. We will offer a generalized version below, so let us dub the above statement, *"De Morgan's Weaker Thesis"*.

De Morgan's defence of his Thesis against Mill's objection[2] that all syllogisms are dependency-circular bears careful scrutiny. The argument is interesting, but lengthy and involved, and we merely sketch the thrust of it here. The objection, writes De Morgan, "tacitly assumes the superfluity of the minor; that is, tacitly assumes we know Plato to be a man, as soon as we know him to be Plato". In other words, Mill's claim that 'Plato is mortal' is part of the evidence for 'All men are mortal' assumes that Plato is a man and not, for example, a dog. But this very assumption is in fact the minor premiss. As De Morgan puts it: "Grant the minor to be superfluous, and no doubt we grant the necessity of connecting the major and the conclusion to be superfluous also. Grant any degree of necessity, or want of necessity, to the minor, and the same is granted to the connection of the major and the conclusion" [50, p. 259]. This is an extraordinarily interesting argument. If correct, it may serve generally to establish that multi-premissed arguments without superfluous premisses never beg the question. Accordingly, we turn now to the question of establishing a condition that will allow us to rule on the question of when a premiss is "superfluous". Of course, in a properly constructed syllogism, there can be no superfluous premiss, but for the contemporary reader the real interest of De Morgan's Thesis lies in the possibility of its application to first-order logic.

4. RELEVANT ARGUMENTS

We propose, following Lehrer [121], that a *relevant deductive argument* is a valid argument in which knowledge of the truth of each and every premiss is required to establish the truth of the conclusion by deduction from the premisses. A *minimally inconsistent set* of statements is an inconsistent set (a set from which a contradiction may be deduced in standard logic), every proper subset of which is consistent. Some inconsistent sets remain inconsistent even when some statement is removed from the set. Others, the minimally inconsistent sets, are such that once any statement is deleted, the remainder is consistent. An argument is a *relevant deductive argument* if the premisses and the denial of the conclusion is a minimally inconsistent set. On the basis of the above pair of definitions, Lehrer shows in [121] that: (i) arguments with contradictory premisses are not relevant, (ii) valid arguments with superfluous premisses are not relevant, and (iii) no argument with a valid conclusion is relevant (or, at any rate, no such argument *with premisses* is relevant). In effect, (i), (ii), and (iii) exclude all arguments except deductively valid ones such that (1) the premisses can be true together, (2) each premiss can be false, and (3) the conclusion can be true and can be false. Hence we have that relevant arguments may contain only contingent statements. On the basis of the foregoing definitions, we now set out the generalized form of De Morgan's Thesis.

De Morgan's Thesis: No relevant argument with more than one premiss begs the question.

We think that this thesis has an interesting degree of initial plausibility, and that it may even be true. There are relevant one-premissed arguments that beg the question, such as at least some of the form, ⌈p, therefore p v q⌉ or ⌈p, therefore p⌉. And there are multi-premissed irrelevant arguments that beg the question, such as ⌈p, q, therefore p⌉. But it is not so clear that there are not relevant multi-premissed arguments that beg the question, given the adequacy of De Morgan's reply to Mill. However, let us look at two possible counter-instances.

5. DISJUNCTIVE SYLLOGISM

One counter-instance to the general formulation of De Morgan's Thesis seems to arise from the possibility of the circular use of the disjunctive syllogism. Suppose some sophist makes it plain that he wishes his victim to infer the disjunctive premiss from the conclusion, in order to evade the obligation of providing independent evidence for that premiss, as schematized below.

(2)
$$\frac{p \vee q \quad \neg p}{q}$$

Transparently evasive, such an argument would perhaps not be very convincing to a sophisticated audience, but if such an argument were proffered, it would clearly constitute a classical *petitio* of the dependency variety. Here we seem to have a counter-instance that evades De Morgan's defence of syllogism. In the syllogism discussed by De Morgan, the remaining premiss was required to establish the fallacy-producing dependency relation, whereas here it is not, since ⌜¬p⌝ is not required to complete the entailment from q to ⌜p ∨ q⌝. Apparently, what we have here is a two-premissed relevant argument that begs the question, contrary to De Morgan's Thesis.

The tenor of De Morgan's remarks indicate quite clearly how he might well have responded to this kind of case, employing what we will call *De Morgan's Defence*, as follows. The counter-instance posits, in effect, not one but two arguments.

(3)
$$\frac{p \vee q \quad \neg p}{q} \quad \nwarrow \quad \frac{q}{p \vee q}$$

The conclusion of the argument on the right is identical to a premiss of the argument on the left. Thus what we have amounts to a combined pair of arguments into a *sequence of arguments*, where the conclusion of the first argument appears as a premiss in the second.

(4)
$$\frac{q}{p \vee q}$$
$$\frac{\neg p}{q}$$

Essentially we have one extended argument with three premisses.

(5)
$$\frac{\begin{array}{c} q \\ p \vee q \\ \neg p \end{array}}{q}$$

But the second and third premisses are superfluous. The second premiss follows deductively from the first, and the third premiss is superfluous for establishing the conclusion, given the first. Indeed, any premiss would here be superfluous, given the first, and in fact (5) is obviously a classical case of straightforwardly logical, explicit equivalence-circularity, De Morgan's preferred type.

The Defence makes an assumption worth stating, namely, that in an array of arguments such as (3) the ones on the right can be absorbed within the ones on the left. The general schema that emerges is that the "evidence" for a premiss can take us back to the conclusion of a previous argument, and that this step may be repeated.

(6)

$$\frac{P_i}{C_i}$$
$$\leftarrow$$
$$\frac{P_3}{C_3}$$
$$\leftarrow$$
$$\frac{P_2}{C_2}$$
$$\leftarrow$$
$$\frac{P_1}{C_1}$$

In any argument, the premisses may drive us back to further argument. The required ruling to ban *petitio* in such a framework is that no C_i must appear as a P_j ($j \geq i$). That is, the conclusion of the initial argument, C_1, must never appear as a premiss, as the argument expands out to the right and upwards.

The prior question that De Morgan's Defence turns on is whether a sequence of arguments, such as (6) above, is an *argument*, or whether it is not, properly speaking, a single argument but a plurality of arguments. To introduce some helpful vocabulary, let is distinguish between an *atomic argument* and a *molecular argument*. An atomic argument is an ordered pair <P,C>, where C is a statement (the conclusion) and P is a set of statements (the premisses). A molecular argument is a finite sequence of atomic arguments, which may also be called the component arguments of a molecular argument. In a molecular argument, a premiss in the component argument may also appear as the conclusion of a previous argument. Given a reasonable-seeming and obvious extension of the notion of relevance to molecular arguments, we have it that no molecular arguments are relevant, for at least one premiss (the conclusion of the next argument to the right) follows deductively from some other premisses (the premisses of that very argument on the right). Thus no molecular argument can ever constitute a counter-example to De Morgan's Thesis. We also have it that if the disjunctive syllogism under-represented by (2) "really" is a molecular argument, it can constitute no threat to De Morgan's Thesis. Thus the general framework suggested by (6) serves to reveal a third thesis that provides a linkage between the Equivalency Reduction Thesis and De Morgan's Thesis: any circular argument, if expanded out far enough to the right, will eventually reveal a premiss identical to the conclusion, and thus will become single-premissed when all superfluous intervening premisses are deleted.[3] The two parts of De Morgan's treatment of *petitio*, although separately perhaps not very plausible, jointly interlock to form a perspective that is both ingenious and deep.

6. CONJUNCTIVE CONCLUSIONS

Another apparent counter-instance to De Morgan's Thesis may be found in an argument of the form below that begs the question.

(7) p
     q
     ―――
     p & q

Arguments of this form need not always beg the question, we suspect, but it is extremely plausible that often they do. Yet such an argument is both relevant and multipremissed. The appropriate strategy for preserving the Thesis against this threat of (7) and its kind might be called *De Morgan's Zwischenzug* (in-between-move): no (single) argument may have a conjunctively molecular conclusion. According to De Morgan's Zwischenzug, (7) becomes the pair of arguments,

(8) $$\frac{\begin{array}{c}p\\q\end{array}}{p} \qquad \frac{\begin{array}{c}p\\q\end{array}}{q}$$

No harm appears to accrue from this move, since of course the pair above are both valid, just as the original two-in-one argument was valid. Some perplexity is generated by the Zwischenzug in other cases, however, for it appears that an argument of the form,

(9) $$\frac{p}{p \ \& \ q}$$

is not simply an invalid argument but really a pair of arguments, one of which is valid (and circular), the other invalid.

(10) $$\frac{p}{p} \qquad \frac{p}{q}$$

This consequence is hard to swallow. We can, and customarily do, tolerate the interchangeability of the following pair of forms,

(11) $$\begin{array}{c}p_1\\p_2\\\cdot\\\cdot\\\cdot\\p_i\\\hline C\end{array} \qquad \frac{p_1 \ \& \ p_2 \ \& \ ...p_i}{C}$$

But the same conjunctive flexibility with respect to the conclusion of an argument must not be allowed. One conclusion per argument is the standard upper limit, and thus it would seem that De Morgan's Zwischenzug leaves logic bereft of (7) and its kind altogether. If it is not feasible to deal with (7) as more than one argument, the only option that remains is to declare it no argument at all. This alternative is not very palatable either, and hence the Zwischenzug is blocked.

Of course, if De Morgan's Thesis were not restricted to multi-premissed arguments, the way out would be obvious. (7) Could be recast as

(12) $$\frac{p \ \& \ q}{p \ \& \ q}$$

And the *petitio* here is manifest. But if the multiple-premiss restriction is dropped, relevant arguments such as ⌜p, therefore p⌝ and ⌜p, therefore p ∨ q⌝ destroy any semblance of plausibility such a strengthened Thesis might seem to possess.

7. CONCLUDING REMARKS

We conclude that De Morgan's Thesis is an extremely interesting object for further study. It would be over-sanguine to suppose that either of the two counter-instances we have developed constitutes a conclusive refutation of the Thesis, but it would be cavalier to dismiss them as easily surmountable. We commend the problem of the confirmation or refutation of De Morgan's Thesis to serious students, if any there be, of that sadly understudied but pedagogically important domain, the informal fallacies.[4]

Let us add that even if De Morgan's Thesis turns out to be true, it would be incorrect, if not fallacious, to conclude that a purely alethic criterion of question-begging would automatically be vindicated. For significantly, the concept of relevance, as we have outlined it following Lehrer, is essentially epistemic. We would also maintain, as we have done in [234], that the weight of evidence supports the hypothesis that the notion of argument appropriate to the *petitio* is essentially epistemic in character. But that is another story.

Chapter 7
Ad Hominem, Contra Gerber

Gerber [75] contains certain proposals regarding the analysis of the informal fallacy, *ad hominem*, on which we shall offer a few comments. We shall set out seven theses which may serve as initially reasonable adequacy conditions of an account of the *ad hominem*, by which we propose to assess Mr. Gerber's own account.

1. DESIDERATA

The conditions here set out involve theses for which, for lack of space, we cannot here argue; but we refer readers to arguments elsewhere in Hamblin [80]. We would also suggest that these are, by and large, supported by the findings of Hamblin [233], [225], [230], [234], and [223].

(1) A fallacy is an incorrect (invalid) argument. An argument is composed of statements. Pure abuse, therefore, does not generally constitute a fallacy. The study of fallacies is not purely a psychological or rhetorical study, but has a logical component as well. A fallacy is not always, or even often, an argument that is first-order invalid or deductively invalid. In other words, the notion of 'correct argument' applicable in the study of fallacies outruns the standard conception of deductive validity.

(2) A fallacy (generally) *seems* correct. Yet not every argument that is, or seems, correct is a fallacy. Insofar as fallacies are systematic errors, particularly persuasive snares and delusions, the study of fallacies must have a psychological component. Hence the study of fallacies could be characterized as "informal" or "applied" logic. There is a danger

here, however, of falling into a naive psychologism and forgetting (1), just above.
(3) *Ad hominem* consists in (i) personal abusive or (ii) various kinds of charges of inconsistency (circumstantial). A constructive approach should reflect this standard orientation.
(4) *Ad hominem* is at least sometimes a fallacious form of argument. It is not always fallacious, however.
(5) It is possible, though not necessarily currently practical, given the present resources of logic, to devise a method for adjudicating between the correct and incorrect forms of *ad hominem*. It is desirable that it should become possible to teach students of logic this method.
(6) Antecedently to (5), it must be possible to recognize or identify the form of arguments *ad hominem* (regardless of their correctness), and to teach students to do so.
(7) There is a special distinction between *ad hominem* and *ad verecundiam*.

Any of these theses might be overruled in the face of contravening developments. However, they reflect what we consider to be a standard view of the *ad hominem* suggested by many, though by no means all, logic texts. They are, in any event, constructive suggestions in that they point to a conception of *ad hominem* that, given the resources of formal logic, seems most likely to be open to explication; and they also function as a practical argumentative device for recognizing avoiding and evaluating particularly pervasive and insidious pitfalls of actual argumentation.

2. NEUTRAL AND ABUSIVE TERMS

Gerber argues [75, p. 24] that bolstering the distinction between the circumstantial and the abusive *ad hominem* requires a detailed taxonomy of abusive as opposed to neutral circumstantial terms. We frankly doubt whether such a taxonomy is even remotely feasible. In other words, we fail to see how Gerber's proposal could ever conceivably meet our condition (5). The three examples offered by Gerber [75, p. 24] make the problem apparent.

Neutral	*Abusive*
large	overweight
restive	insubordinate
killer	murderer

As for the first two pairs, it is not easy to see how there is any significant semantic connection between the first term and the second. It is well known that a macrosomatous individual need not be overweight, and conversely. Similarly, the lack of any *prima facie* semantical connection between restiveness and insubordination makes it hard to see how the first is "neutral" in some identifiable respect in which the second is "abusive". The semantical connection evidently posited between the third pair might perhaps reside in an equivalence that could be crudely put as follows:

a murders *b* if *a* kills *b* intentionally

Here we might understand the adverb on the right as a gloss for the heterogeneous variety of defences recognized under *mens rea*. Perhaps then, Gerber means to rule that the abuse resides in the allegation of *intention*. This is a kind of claim commonly seen in writings on the fallacies, that the fallacy depends on the intent to deceive. But it is a claim that, following (1), we reject. Fallacies are not purely psychological, and if fallacy F_1 is a distinct fallacy from F_2, then F_1 should be an instance of a form of incorrect argument distinct from F_2. We hasten to add that it is by no means clear that Gerber wishes to appeal to intentions. The problem is that it is simply not clear *what* criterion he wishes to appeal to.

To postulate that what is or is not a fallacy turns on an "abusive-neutral" dichotomy seems to us a questionable psychologism which it is best not to acquiesce in and which condition (1) counsels against. Strangely enough, Gerber seems to recognize this subjective aspect himself, writing [75, p. 25] that the same term can be variously laudatory, neutral or condemnatory, depending on the features of the situations in which it is employed. The distinction, he writes, is a "function of the occasion of their use". Perhaps enough has been said to dissuade even the most optimistic reader that the Gerber project for the analysis of *ad hominem* is started in the right direction.

3. LAUDATORY AD HOMINEM

Gerber suggests that one form of *ad hominem* is the scheme we will call S_1.

a is P
a says (or believes) that q
Therefore, q is false (or true)

One possibility among the various correct argument forms suggested by the above scheme is the laudatory *ad hominem* argument. An instance: *a* is genuine authority in a domain *d*, and *a* says that q (where q is an assertion relative to *d*), therefore q is (probably) true. Contrary to Gerber, and in accordance with (7), we think that this latter form of argument is that of the *ad verecundiam*, specifically the appeal to expertise see [167] and [225]. But we agree with Gerber that there is a significant and interesting connection between the *ad hominem* and (what we call) the *ad verecundiam* argument.

The deficiency of S_1 as an account of the *ad hominem* lies in its over-generality. Clearly, only certain properties can stand in for P in a genuine *ad hominem*. We maintain that Gerber simply fails to provide a workable set of conditions on the P-class. At any rate, it is quite clear that S_1, by itself, does not provide an analysis of the *ad hominem* or *ad verecundiam*.

4. LOCKEAN AD HOMINEM

John Locke described the *ad hominem* as a mode of refutation or assent-mongering by "pressing a man with the consequences drawn from his own principles or concessions" [125, p. 278f]. Locke contrasted *ad hominem* with "the using of proofs drawn from any of the foundations of logic or probability". Hamblin notes in [80] that the Lockean distinction is very close to the distinction that Aristotle drew between "dialectical" and "demonstrative" arguments. On the Lockean view, of course, *ad hominem* is not necessarily fallacious, for we may assume that it is quite often appropriate, reasonable, and logically unexceptionable to press a man with undesirable or unforeseen logical consequences of his assertions or beliefs.

How then, on the Lockean view, can *ad hominem* be a fallacy at all? Certainly arguing from an adversary's premiss is a procedure that could be abused, or excessively indulged in, or confused with an appeal to "independent evidence", where the latter is available. But it is difficult to see how these abuses are to be specified. Gerber is not much help here. He suggests, in a Lockean spirit, that the class of circumstantial *ad hominem*, presumably distinguishable from S_1, is captured by the scheme S_2:

a says or believes that q
r is a logical consequence of q
Therefore r is true

Gerber points out that there can be both abusive and circumstantial varieties of this scheme, and also a laudative variety. As with S_1, he urges that both valid and invalid instances of S_2 are possible. The problem is that, aside from more appeals to his "abusive-neutral-laudatory" taxonomy, Gerber does not provide a set of conditions for adjudication. The general problem with the Lockean view remains. It is simply not clear why or when arguing from an opponent's premiss is fallacious. Gerber is tickled by the discovery that there are arguments that fit the qualifications for *ad hominem* and are perfectly valid [75, p. 27]. However, this discovery suggests to us merely that S_2 is too broad an account of *ad hominem* to meet the adequacy condition (3).

5. CONCLUDING REMARKS

Many texts imply that the *ad hominem* is always fallacious. They do not appear to recognize that a comprehensive theory that would explain the fallaciousness of the oft-cited examples of *ad hominem* is also a theory in which it is likely that there are *valid* forms of argument pertaining essentially to the character, circumstances, or situation of the arguer. We think that the articulation of this point is a valuable feature of Gerber's paper. We have argued elsewhere in [233] that it is helpful to recognize that the informal fallacies often conceal *correct* forms of argument, even if the required concept of correct argument outruns standard notions of deductive validity. This much we can happily accept. But we do not think that the direction taken in Gerber's suggestions for the mechanisms required to sort between the correct and incorrect cases is a felicitous or constructive one. Nor can we bring ourselves to think that Gerber has offered any good reasons for classifying what we consider the classical *ad verecundiam* with the *ad hominem*, in violation of adequacy condition (7). Of course, it would be disastrous to accept all the textbook banalities about fallacies, but they do set, in some important respects, a standard and tradition that is closely bound to the pitfalls of actual argumentation that is therefore worth preserving, within the constraints of the theory of fallacies itself, as it develops. On the standard view, what Gerber offers as an example of *ad hominem*,

> Jones is an infallible mathematician
> Jones says sincerely that $e^{\pi i} = -1$
> Therefore, $e^{\pi i} = -1$,

is not an *ad hominem*. Aside from certain reservation expressed in [225], it is a paradigm of the classical *ad verecundiam*.

Chapter 8
Composition and Division

The informal fallacies make for an unsatisfactory sector of logic. Other branches of logic have in recent times achieved an impressive theoretical development, but accounts of the fallacies have changed except for their banality very little since Aristotle. Not that fallacies are uninteresting. Students of logic often exhibit a lively initial interest in the fallacies, of which the continued appearance in textbooks attests to their practical importance in actual argumentation. Characteristically, however, the interest is short-lived. The lack of clear and adequate characterizations makes it virtually impossible to commend fallacy-theoretic notions to students as an effective strategy in the analysis of argumentation. Ironically, forms of argument identified as "fallacious" by the textbooks often turn out to be correct arguments, even if the salient concept of correctness is somewhat theoretically elusive.

We direct our efforts here to understanding something of the theoretical basis of the fallacies of composition and division. Our analysis will ultimately provide that these two fallacies are, as the tradition teaches, virtually two sides of the same counterfeit coin. So from time to time we will also refer to them singularly as 'the fallacy.'

1. THE MEDIAEVAL VIEW

§1. The mediaeval view seems to derive most of its inspiration from the brief account of composition and division in the *De Sophisticis Elenchis* [8], 166a 22-166a 37. Aristotle gives four classical examples, three of which bear remarking upon. (1) The expression 'A man can walk while sitting' is true when interpreted in *a divided* sense, as saying that the sitting man has the power to walk. But the same sentence is false if the words are *combined*, so as to say that it is possible for a man to walk-while-sitting. It may well have seemed to Aristotle that this fallacy was purely syntactic, and hence "inside language", because it can be eliminated by

94   Chapter 8

making explicit the scope of what nowadays we would call the modal operator. The sentence

(I1) $(M\vartheta \ \& \ M\psi) \rightarrow M(\vartheta \ \& \ \psi)$

is not a theorem in the standard modal logics, and it may seem that the fallacy is easily resolved by the appropriate rearrangement of parentheses. But it could also be supposed that, rather more deeply understood, the fallacy is a systematic kind of equivocation on the word 'possible.' We can see the impossibility of a man both walking and sitting at t. But when we say that a man has the power to walk at t, even though he is not actually walking at t, we might mean that it is possible relative to conditions *before* t, for him to walk at t, for if he does not walk at t, then it is not possible for him to walk at t, relative to conditions at t, Perhaps this underlying issue is the reason why the Kneales [114, p. 93], argue that the distinction between absolute and relative possibility is implicit in this historically influential example.[1] So, more than merely syntactical considerations could be involved. Still, if the fallacy is an equivocation on 'possible,' it is nonetheless a fallacy within language, since equivocation is a verbal fallacy.

(2) A second example plainly enough resembles (1): 'A man can write while not writing.' can mean that when a man is not writing he has the power to write (sometimes true) or that a man has the power to both write and not write at once (always false).

(3) The third example is: "He knows now if he has learnt his letters." (166a 30). This is somewhat similar to an example, which we presently take up, from the *Rhetoric*. Here the fallacy, which is at least partly linguistic, may be elucidated as follows. 'Knows' is evidently meant here in the sense of 'understands.' If we read '$U_a\vartheta$' as 'a understands the expression ϑ,' then consider

(I2) $(U_a \ \vartheta \ \& \ U_a \ \psi) \rightarrow U_a(\vartheta \ \& \ \psi)$.

Again, the fallacy is, after the fashion of (1), a question of mismanaged modal scope.

Of division, three notable examples are offered by Aristotle. The first two are mathematical. (1) 5 is 2 and 3, and therefore even and odd. (2) The greater is equal, for it is that amount and more besides. The first inference is a fallacy because although it is true that 2 and 3, taken separately, are respectively even and odd, it is false that the sum, 2+3, is both even and odd. More fully, the argument could be set out:

(A1) P1: 5 is 2 and 3.
      P2: 2 and 3 are even and odd.
      C: Therefore, 5 is even and odd.

P2 is (at least) three-ways syntactically amphibolous:

(P2a) (2 and 3) is even & (2 and 3) is odd.
(P2b) 2 is even & 3 is odd.
(P2c) 2 is even & 2 is odd & 3 is even & 3 is odd.

though, to be sure, (P2c) is not very plausible. The likelier source of the sophism is the conflation between the true (P2b) and the false (P2a). (P2a) is false if we interpret the 'and' as a functor for the operation of addition, since, accordingly, the left conjunct is false. As Hamblin observes in [80, p. 20], however, perhaps there is also an equivocation involved, for it seems natural to interpret the 'ands' in P2 as expressing conjunction and the 'and' in P1 as expressing the arithmetical operation of addition.

In (2), the amphibolous conflation appears to be between the truth that $x > y. \supset .x = (y + z)$, and some false analogue such as that $x > y. \supset .(x = y) + z$. Here too there could be an element of equivocation on 'and'. Both examples illustrate how the cases of composition and division which Aristotle dealt with in the *De Sophisticis Elenchis* tend to reduce to cases of amphiboly or equivocation, and are thus correctly enough said to be linguistic fallacies.

(3) The third fallacy is likewise linguistic: 'I made thee a slave once a free man.' (166a 36). This could mean 'You who were once a slave I made a free man' (divided) or 'I made you into a person who was once a free man and is now a slave' (composed, i.e., I made thee a-slave-once-a-free-man). Again this is a matter of scope. If we were to think of "bringing-about" as a modal operator (see §6) and Walton [66], [194]) the critical difference is that between the pair:

(a brings it about that p) & q
a brings it about that (p & q)

So here too amphiboly seems to be a significant factor.

§2. From these examples in the *De Sophisticis Elenchis* arose the Scholastic doctrine of the composite and divided senses of propositions. Peter of Spain [144] gives examples (1) and (2) of composition and analyses them in somewhat the same way as we have here. Curiously, the 'and' in Aristotle becomes 'or' in Bochenski's translation of Peter's example, e.g. 'five is even or odd.'[2] This leads Bochenski to describe Peter's discussion of an interesting example exidently also discussed by Burleigh (both texts are translated in Bochenski [24, p. 186f.]. Burleigh considers this argument,

(A2)   Every animal is rational or irrational.
       Not every animal is rational.
       Therefore, every animal is irrational.[3]

This is a sophism, writes Burleigh, since the conclusion is false, but the minor is not false. Therefore the major must be false. But this is absurd since surely the major is true. Burleigh's solution is that the major is said to be "multiple, according to composition and division". In modern terms, as Bochenski adds in a note [24, p. 187], the difference is simply that between

$(\forall x)$ (x is an animal $\supset$ (x is rational $\vee$ x is irrational)
and
$(\forall x)$ (x is an animal $\supset$ x is rational) $\vee$
$(\forall y)$ (y is an animal $\supset$ y is irrational),
although both Peter and Burleigh evidently have in mind
$(\forall x)$ (x is an animal $\supset$ x is (rational or irrational)).

Peter, among others, asserts that what we call "disjunction" can "disjoin" either terms or propositions. So it is intriguing to read that disjunction as well as conjunction can be a source of composition and division. Later we will take up these kinds of fallacy in more detail.

§3. William of Sherwood, in [210, p. 140], makes it plain at theoutset that in speaking of composition and division, he is speaking of connection and separation in the act of speaking (*in actu proferendi*). In a footnote, Kretzmann adds that the mode of dividing or compounding in spoken discourse is often by *pausing* or *emphasizing*. This observation shows how composition and division are connected with yet another fallacy within language, namely *accent*.

Kretzmann adds in the same note: "The Aristotelian-mediaeval doctrine of the fallacies of composition and division is much broader than that found in textbooks of the modern period, which usually recognize only the difficulties associated with the compounded and divided senses of generalizations, such as 'all the interior angles of a triangle equal 180°'" [210, p. 140]. We might call this *Kretzmann's Thesis* concerning the relation of the mediaeval to the modern view of the composition and division. William's remark about acts of speaking is extremely interesting, and it suggests the plausible thesis that the mediaeval doctrine was predominantly concerned with linguistic instances of these fallacies. In contrast, the modern period would seem more concerned with less linguistically dependent examples, having to do notably with parts and wholes. (Though in *Rhetoric* 1401a, Aristotle also deals with such cases under the heading of Division and Composition.)

William's treatment is very much in the spirit of the *De Sophisticis Elenchis*. He offers an interesting example

(A3)   Whatever is possible will be true.
       That a white thing is black is possible.
       Therefore, that a white thing is black will be true.

Without dwelling on the analysis of this example, suffice it to say that it is treated very similarly to Aristotle's examples (1) and (2) of composition. William also offers Aristotle's (1) and (2) of division, nearly verbatim, and some interesting tensed examples are given as well [210, p. 142ff.]; *vide* also [211]. In many ways William's treatment is typical of the Scholastic conception of composition and division, particularly in its emphasis on linguistic collocations and separations. In contrast, the modern conception of these fallacies is generally non-linguistic and seemingly more "substantive" in its preoccupations. This contrast is partially misleading as we will see later, but for now let us turn to a consideration of the modern view.

2. THE MODERN VIEW

§4. According to a prevalent perspective, which we dub *the modern view*, quintessentially exemplified by Copi [40], the fallacies of composition and division have to do exclusively with the part-whole relation or the notions of distributions and collections (*physical* parts and wholes are mainly meant here). Accordingly, there are two varieties of the pair of fallacies: (1) The fallacy of composition$_1$ is committed when it is argued that a whole has a certain property simply because all its parts have that property, e.g. "A particularly flagrant example would be to argue that since every part of a certain machine is light in weight, the machine "as a whole" is light in weight" [40, p. 96]. The fallacy of division$_1$ is the converse of this. (2) The fallacy of composition$_2$

is committed when it is argued that a totality has a certain property simply because each of its members has a certain property, e.g. ".... it would be fallacious to argue that because a bus uses more gasoline than an automobile, therefore all buses use more gasoline than all automobiles" [40, p. 96]. In the traditional terminology, buses taken *distributively* use more gasoline than cars taken distributively; i.e. for all x, if x is a bus then if for all y, y is a car, then x uses more gas than y. But *collectively*, cars use more gas because there are more of them. Likewise, division, is the converse fallacy. The difference between (1) and (2) resides, presumably, in the circumstance that a machine is an actual physical composite of its parts and a functional composite even when disassembled, whereas the "totality of cars" is a definitional collocation thrown together by "an act of the mind". As Copi puts it, "A whole like a machine, a house, or a wall has its parts organized or arranged in certain definite ways" [40, p. 97]. Hence there is a distinction to be made between an "organized whole" and a "mere collection". By the latter term, however, Copi evidently does not mean *set*, since the set of buses does not use gasoline. In part V we return to the matter of what we there call "aggregates".

§5. Much of the inspiration for the modern view could well have ultimately been derived from *Rhetorica* (1) 1401a 24-1401b 30, where Aristotle discusses some sophisms of arguing from the parts to the whole, and conversely. Three of Aristotle's examples, in particular, are of interest. (1) The argument that ".... one who knows the letters knows the whole word, since the word is the same thing as the letters which compose it ..." (1401a 29). This is somewhat similar to Copi's example of the machine insofar as the word is a functional construct of its letters. Yet the relation involved here is different, unless we take 'word' merely at the phonological or orthographic (that is, physical) level. In fact, the sort of part-whole relation here involved could be a very difficult question to analyze. (2) ".... if a double portion of a certain thing is harmful to health, then a single portion must not be called wholesome, since it is absurd that two good things should make one bad thing". This example seems to turn simply on the matter of physical division independently of functional relationships. (3) According to an argument in the *Orestes* of Theodectes, it is right that she who slays her husband should die, and it is right that a son should avenge his father. Jointly, as in the case of Orestes, perhaps they are a less than allowable act. Distributively they are right, but collectively not.

Hamblin [80, p. 84], observes that the account in the *Rhetoric* has moved completely away from the one in the *De Sophisticis Elenchis*; and, as we have just seen, the account in the *Rhetoric* does bear a significant resemblance to the modern view. It is worth remembering, however, that there is also a verbal aspect to (1) and (3) evocative of the mediaeval view of the *De Sophisticis Elenchis*. This element is obvious in (1). It would also be obvious in (3) to someone familiar with the syntax of deontic modal logic (not Aristotle), for on that analysis, the fallacy would syntactically resemble the walking-sitting argument of the *De Sophisticis Elenchis*. What is being suggested is that the following inference is fallacious:

(I3) (p is right & q is right) → (p & q) is right.

Yet there is a significant disanalogy revealed by a deeper analysis. The real sophism resides rather in this inference:

(A4) p is right for **a**
  q is right for **b**
  Therefore, p & q is right for **a** = **b**.

(I3) has to do with the scope of the operator, and so is syntactical (within language). But it is difficult to see how (A4) is likewise syntactical. One might make similar kinds of comments about (1), although it is ostensibly linguistic, at a deeper level it seems less a syntactic than a substantive affair, and so is rather more akin to the modern view than the mediaeval.

3. THE TWO VIEWS JUXTAPOSED

§6. What appears to be a clear contrast between the mediaeval and modern conceptions of the fallacies of composition and division may now be ventured: the mediaeval view pertains to linguistic collocations and separations, and is firmly within the tradition of the account in the *De Sophisticis Elenchis*. The modern view pertains to (a) extralinguistic physical parts and wholes and (b) the collective-distributive distinction and is more in the tradition of the *Rhetoric*. Accurate up to a point, the contrast is blunted by the fact that the Scholastics were also very clearly aware of (a) and (b) and treat of both areas extensively. But here, we think, is the source of confusion: it is not explicitly clear that the account in the *Rhetoric* is meant by Aristotle to be an account of *the fallacies of composition and division*. Moreover, the mediaeval accounts of (a) and (b) are traditionally classified separately from the topics of composition and division. Over time a name-change has taken place. What was called composition and division by Aristotle and the mediaevals have largely been forgotten and the names "compositions" and "division" have in modern times been appropriated for the traditional mediaeval writings stemming from the *Rhetoric*, these having to do with (a) and (b). Thus *Kretzmann's Thesis* is confirmed. The mediaeval doctrine is, in a sense, broader than the modern. We have had, on the face of it, two quite distinct developments, as below:

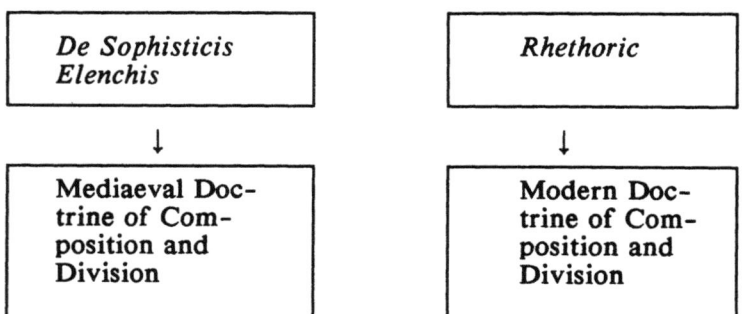

where the *Rhetoric* bypasses the mediaeval period altogether. But, more accurately, there is a continuity disguised by the shift of terminology.

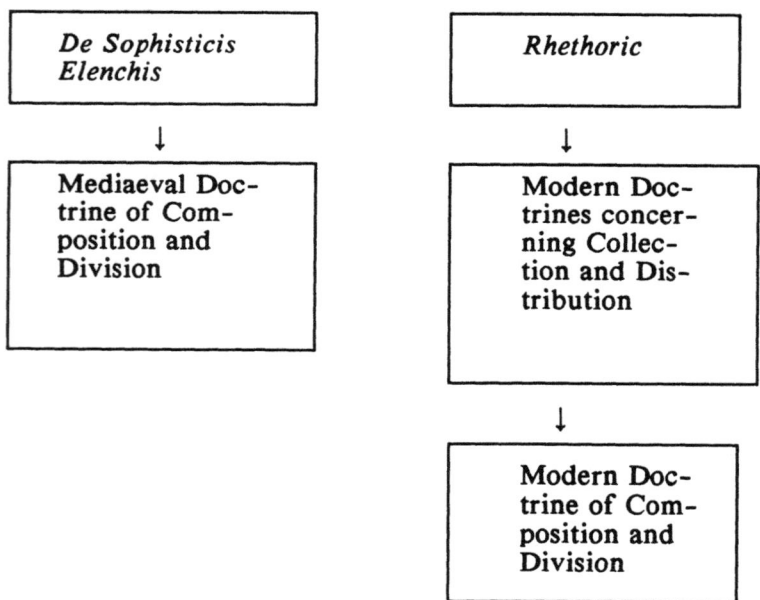

Here, the development of the left has largely atrophied and is not carried over into modern times very significantly. We still have two separate developments, but if we think of the two mediaeval boxes as joined, and the Aristotelian boxes as joined, then Kretzmann is correct to say that the Mediaeval-Aristotelian doctrine is broader than the modern. Of course it could also be surmised that there have simply been too separate streams of development and that the mediaeval doctrine is different from the modern one, not broader. But now that the larger picture shows itself, one can see that either view can give rise to confusion, as witness many a modern text and commentary. One such confusion is reflected in the feeling of extreme puzzlement among many commentators that Aristotle or the Scholastics should have classified composition and division as fallacies within language at all.

William of Sherwood's treatment of (a) and (b) is especially instructive [211, p. 39]. He distinguishes very clearly between 'every' or 'all' taken distributively or collectively and he also shows himself familiar with modern-sounding *sophismata* having to do with parts and wholes [211, p. 40]; yet makes no mention of composition or division in this connection.

§7. In reconstructing an analysis of composition and division adequate to our current needs and consonant with developments in other branches of logic, we are now faced with a choice. Should we leave the mediaeval view dormant, as the modern texts typically do, or should we strive to join the two views into a unified whole? We will now introduce some concepts which suggest, we think, that the latter course is at least feasible. Later we will face the question of which seems to be the better course, assuming both of them to be open.

4. BASIS OF THE FALLACY: AGGREGATES

In searching out the logical structure of composition and division, it is useful to bear in mind that is *as fallacies* that they command our attention as logicians. "A mistake in reasoning from whole-to-parts or from parts-to-whole". But what kind of mistake? If one takes the essence of the fallacy to be that of inferring to what should not *conceivably* be the case, as when inferring from the premiss that a set of apples is abstract that its member apples are abstract, then the principal transgression would be modal, evocative perhaps of what used to be called "category mistakes". In that case, a decent theory of the compositional and divisional fallacies would need to incorporate a decent theory of categories or kinds.

However, there may be reason to think that not every such fallacy is of the "it could not possibly be so" sort. If I argue that, since the Montreal Canadiens are a great hockey team each player is a great player, then I make a mistake. The conclusion is untrue in fact, not as a matter of necessity. My argument is *deductively invalid*, and that may be all that there is to the *logic* of these fallacies: a compositional or divisional argument is a fallacy just when the truth of its premisses is compatible with the falsity of its conclusion. Or, it may be that a rather more interesting locus of these fallacies is to be found in defective (and suppressed) *inductive* linkages, as in: "The Montreal Canadiens are a great team, ergo all the players are great players", where there is a kind of inductive mediation as follows. Since the team is great, *most* players are great (fair enough), and if most are, all are (the induction). A stupid induction to be sure. For if the strongest allowable immediate inference is that most are great, then it is anticipated that some are not; yet it is precisely this anticipation that is overridden by the suppressed induction.

Category mistake, deductive invalidity, inductive infelicity, whatever the logical scaffolding might be, it is plain that a large part of the more interesting story is to be found elsewhere, whether in the semantic structure of the predicates involved, or in the notion of part that underlies compositional and divisional configuration. In the present essay, we shall concentrate our efforts on the latter notion, and leave the former for another occasion.

§8. Composition and division, fallacious or not, are a motley of practices transacted, soundly or not, in myriad settings having to do loosely with relationships of parts-to-whole and whole-to-parts. Such transactions are influenced by a host of intuitive distinctions by which, however crudely, the distinction between allowable and non-allowable composition and division are approximated to and illuminated. Examples of these distinctions spring readily to mind.

(i) Objects specified via their "natural" *properties* vs. objects specified via their *conventional or functional properties*. *Example*: Though the wooden thing in the corner = the desk in the corner, division holds only with respect to being a wooden thing.

(ii) Properties of an object *denoted by adjectives* vs. properties of an object *denoted by nouns*. *Example*: Though the red outfit in the closest = the (two-piece) bikini in the closet, division holds only with respect to being red.

(iii) Properties of an object that are *non-relative* vs. properties of an object that are *relative*. *Example*: Though the wooden thing over there = the heavy thing over there, division holds only with respect to being wooden.

(iv) The *quality* of a thing vs. its *quantity*. *Example*: Though the red outfit = the bikini-shaped outfit, division holds only with respect to being red.

(v) The *secondary qualities* of a thing vs. its *tertiary qualities*. *Example*: Though the coloured mosaic = the beautiful thing over there, division holds only with respect to being coloured.

It takes little effort to convince oneself that composition and division verdicts based on such intuitive distinctions as these are not very satisfactory; in particular, they have a way of not generalizing. Though the lemonade is yellow (secondary quality) and tasty (tertiary quality), it is far from clear that all its parts are yellow and not all tasty. It depends on what is here meant by parts. If *being a desk* does not divide, deskhood being a conventional or functional property, what do we say of *being a Chinese box*, also a conventional or functional property? For an utterly natural construal of 'divides,' a Chinese box divides only into Chinese boxes (charity commanding perhaps that the terminal remainder be a limiting case of Chinese boxhood). So, again, it depends on what parts are.

Parts are not only difficult to specify or define in some suitably general fashion. They also make trouble for attributes of action. Thus, if I *use* the hammer, do I use all its parts? If I am driving a nail, I use the hammer to drive but the head-face to strike, the handle to grip, etc. So there is the question of whether even in such "divisions" as these, a single selfsame attribute is being divided. And what, too, of the claws of the head; surely, though the hammer's head is used in driving, the claw-parts are not, or not certainly in the same sense of "use". If you kick Gregor in the rump you kick Gregor; and if you kick Gregor you kick him somewhere (i.e. some part of him). Bit what if you kick Gregor's sole strand of hair (worn exotically, as a seven-foot strand), do you kick Gregor? Do you kick a Gregor-part? It depends in what one count as a part.

It is necessary, therefore, to theorize about relations of part-to-whole. To contemporary ears this sounds like a charge to do the theory of sets. But set theory gives an explication of parts and wholes of virtually no interest to the logic of composition and division. Sets are not sited in space-time, do not admit of being acted upon or perceived. A set of dominoes cannot be played with, a set of good pearls cannot be worn or mislaid, and the read two pieces of Xantippe's bikini do not constitute a red set. Sets are required to be abstract; therefore where the members of sets are concrete, most interesting composition and division questions are answered automatically in the negative. The way of sets is too facile for our purposes.

A better choice than set theory for the reconstruction of the part-whole relation would be something like mereology in the fashion of Lesniewski [123]. For one thing, a significant aspect of the motivation of mereology was to hit upon a theory of classes, not only as actual collections of their objects, but also as collectivities which would mirror some of the features of collective as opposed to distributive predication. An interesting development of what is basically a mereological theory can be found in Noll's contributions to a theory of bodies adequate for the foundation of mechanics, and in an extended and refined version of Noll's account, due to Suppes [181, p. 392-5], a version which we shall follow here. We need a workable notion of part in order to plumb the logical structure of composition and division. Bodies, paradigmatically perhaps, have parts, and so we may expect a theory of bodies to say something useful about parts.

In the Noll-Suppes theory 'π' can be taken to designate the part relation. The following definitions ensue:

If $B \pi A$ and $C \pi A$ then A is an *envelope* of the pair-set $\{B, C\}$. A is the *least envelope* of $\{B, C\}$ if, and only if, A is an envelope of $\{B, C\}$ and for any X that is an envelope of $\{B, C\}$, $A \pi X$.

A is a common part of $\{B, C\}$ if, and only if, $A \pi B$ and $A \pi C$. If A and B are bodies, they are separate, if, and only if, they lack a common part. If A and B are bodies, then A is a least part of B if, and only if, $A \pi B$ and there is no body C such that $C \pi A$ and $C \neq A$. Body A is the *greatest common part* of $\{B, C\}$ if, and only if, A is a common part of $\{B, C\}$ and for every body X, if X is a common part of $\{B, C\}$, $X \pi A$.

Furthermore, if A is the least envelope of $\{B, C\}$, then the set-theoretic union of B and C is identical to A; A, then, is the *join* of B and C. If A is the greatest common part of $\{B, C\}$, then the set-theoretic intersection of B and C is identical to A; A, then, is the *meet* of B and C (The operations of join and meet are partial).

If $A_1,..., A_n$ are parts of B_1, if $A_1 \cup ... \cup A_n$ exists, and if $A_1 \cup ... \cup A_n = B$, then $\{A_1,...,A_n\}$ is a *finite dissection* of B.

Finally, a *binary structure* W = <W,π> is a *structure of bodies* if, and only if, the following axioms hold for all A, B, C and D in W:

1. $A \pi A$. (Reflexivity)
2. If $A \pi B$ and $B \pi A$, then $A = B$. (Asymmetry)
3. If $A \pi B$ and $B \pi C$, then $A \pi B$. (Transitivity)
(Hence the part-relation gives a partial ordering of the bodies in W).
4. If A and B have a common part, then they have a greatest common part.
5. If A and B have an envelope, then they have a least envelope.
6. If A is a part of B and $A \neq B$, then there is a body C in W such that B is the least envelope of $\{A, C\}$.
7. Every body has a least part.
8. Every body has a finite dissection of least parts.
(Note that axioms 7 and 8 commit the theory of bodies to atomism; by 7 every body contains at least one atom, and by 8 every body is composed of finitely many atoms.)

The Noll-Suppes explication of part is evidently better-suited to our present concern that the standard idea of set-membership. But it is still a trifle heavy-handed. Suppes, mind you, has different objectives in mind. He seeks for (i) a notion of body appropriate for a geometry of solids with a richness of structure equivalent to Euclidian geometry; and (ii) an explication of a non-container concept of space via, not points, but (bodily) positions.

We shall confine ourselves to three examples. The axioms for the part-relation provide that whenever a property F holds of all the parts of a whole (i.e. body), it holds of the whole, that is, that properties true of all parts *always* compose. On the contrary, however, what makes composition so interesting to a logician is that it is sometimes a fallacy, the fallacy of supposing that what is true of all parts is true of the whole. The Noll-Suppes notion of part entails that composition is not a fallacy; and that, we say, is too heavy-handed a consequence. Not that this comes as any surprise to a Suppesian body-theorist. *Any* theory in which the part-relation is reflexive is a theory which automatically provides an affirmative answer to the question, "If all the parts have F, does the whole likewise have F?". Such theories are of no interest for the analysis of composition; however, an easily restricted version of it goes easily back into contention. Since it is in the nature of improper, parts always to compose, the right restriction to make is to proper parts, whereupon the reflexivity axiom is retired. However, in the atomism of this theory (restricted version or not) there lie other unwelcome implications. Intuitively, if a chain is pure iron so are its parts. But the Noll-Suppes theory of parts does not require that the smallest parts of a chain be its links. It leaves it entirely open for whatever further part-dis-sections a good physical-chemical theory of iron links would provide. In particular, the Noll-Suppes theory allows the *chemical* atoms of the links to be parts of the chain. In consequence, a surprising number of properties which, intuitively, we would expect to compose and divide fail to do so; and fallacies are attributed where none there are.

It is useful to emphasize that the very idea, e.g. that division is sometimes *not* a fallacy directly challenges a philosophical well-known, perhaps even entrenched, view of parts. It can be called without violence to the facts a "Leibnizean" conception of parthood, and it provides that every member of the parthood descendent class of a whole, A, is itself also a part of A. In particular, such a view implies that chemical atoms and links are alike *bona fide* parts of a chain, and, therefore, that the predicate "is made of iron" does *not* divide with respect to an iron chain. However, we shall not here seriously pursue the question of the legitimacy of "Leibnizean" parthood. Suffice it to say that it would have unwelcome consequences for the theory of the fallacies presently before us.

A third difficulty is this. Nothing in the Noll-Suppes theory calls for restrictions on predicates of bodies. From the point of view of an account of the fallacies of composition and division, this requires correction. Suppose, for example, that, if F is a property that composes (or does not) in the whole W, then any property entailed by F likewise there composes (or does not), i.e. that composition and non-composition are closed under consequence. Suppose further the existence of the tautological property, T, i.e. of a property which holds of everything whatever. Then for any predicate, F, F is entailed by the disjunctive property '$F \vee T$'. Let F be a property which, intuitively, does *not* compose in a whole W. But '$F \vee T$' does composes there; hence by closure under consequence so does F, never mind our intuitions.

Obviously, various technical remedies are open to us, including the rejection of the closure property. However, closure could be retained if we restricted our theory of composition and division to what Woods in [217] has called "semantic kind properties".

This would have the consequence of furnishing the theory of composition and division with a large and rich array of properties without having to trouble with such unprojectibilia as "F or T" and other sorts of "unnaturally" complex attributes.

Our final pass at a theory of parts will require us to introduce, and take seriously, the idea of an *aggregate*. The notion we here present is essentially that of Tyler Burge in [28]. For current purposes, aggregates can be linked to first-order sets, i.e. sets containing only individuals as members. However, as shortly will become clear, aggregates are significantly different from sets. For example, we will put it that the empty set is *not* an aggregate and that no singleton is an aggregate; from the point of view of aggregate-theory, a singleton of which the sole member is an individual just is that individual. As will shortly become clear, aggregates also exhibit useful differences from mereological classes and Suppesian bodies. We turn now to the formal development. The base logic for this theory is straightforward. See Burge [29].

In place of the set-theoretic notion of membership, given by the ε relation, aggregate theory speaks of *componentiation*, i.e. being a component-member, symbolized by 'a.' Wherever x bears to y the a-relation, we could say, somewhat barbarically, that x *componentiates* y. The predicate expression, 'ax' for 'is an aggregate,' can be defined thus:

(i) $ax \leftrightarrow (\exists z)(\exists y)(zax \wedge yax \wedge z \neq y)$

That is, x is an aggregate if, and only if, at least two objects are component members of it. A definition is also ready to hand for aggregate-abstraction:

(ii) $\hat{x}/\vartheta x = df\ (\forall y)((\forall x)(xay \leftrightarrow \exists z)(\vartheta z \wedge xaz))$

The counterpart of set-comprehension is:

(iii) $(ya\hat{x}/\vartheta x) \leftrightarrow (\exists z)(\vartheta z \wedge yaz)$

An explicit denial of the empty aggregate can be got from:

(iv) $\neg(\exists y)(y = \hat{x}/x \neq x)$

Analogous to the principle of extensionality for sets we have:

(v) $(az \wedge ay) \rightarrow (y = z \leftrightarrow (\forall x)(xay \leftrightarrow xaz))$

where the antecedent reflects the intention of the theory that not everything be an aggregate. In fact, our next principle makes the point more precisely: only individuals are components of aggregates.

(vi) $xay \rightarrow Ix$,

where individuals may be taken to be their own components:

(vii) $Ix \rightarrow xax \wedge (\forall z)(zax \rightarrow z = x)$

The notion of an *individual* is defined by

(viii) $I(x) \leftrightarrow (\exists w)(xaW)$

It is clear at once that the a-relation also significantly differs from the π-relation of the Noll-Suppes theory of bodies. Whereas, $A\ \pi\ A$ holds, $a\ a\ a$ does not (for A a body and a an aggregate). Moreover, not all parts of an aggregate are components of it. The transitivity principle, if $x\ a\ y$ and $y\ a\ z$, then $x\ a\ z$, does not obtain where y and z are aggregates. However, transitivity is provable in case x and y are just the same individual. In fact, given the joint assumption of $x\ a\ y$ and $y\ a\ z$, it *follows* that $x = y$, and therefrom that $x\ a\ z$. The proof may be found in Appendix A. So aggregates are not mereological sets or Suppesian bodies. Aggregates are entities that suggest themselves for the analysis of idioms in which there are plural constructions and mass terms, e.g. for such expressions as "the stars that presently made up the Pleiades galactic construction". Unlike sets, aggregates are physical entities in space-time, capable of action and of change, and susceptible of coming into and going out of existence. Unlike mereological classes and Suppesian bodies, not all parts of an aggregate are parts that *make up* the aggregate; no aggregate is its own member-component; and no aggregate may be the component member of any aggregate (i.e. aggregates are always aggregates of individuals). By virtue of the first of these three differences, one might be able to think of an iron chain as an aggregate, the ironness of which divides over its parts (i.e. its member-components, i.e. its links), and not worry about, e.g. the atomic parts of these components. They are parts that do not matter.[4a] And by virtue of the third of these three differences, aggregate theory avoids Russell-Zermelo problems of aggregates of all aggregates that are not component-members of themselves. In Burge's theory, predicates of aggregates are kept reasonably well-behaved and projectible. Consequently, much the same purpose is served as would be by a restriction of properties to semantic kind properties – either way, one need not trouble with such mischievious disjunctive properties as "*F* or *T*" Aggregate-theory is rather well-suited, therefore, to our interest in composition and division. All the same, the obvious relevance of aggregates to the logical theory of composition and division should not be over-estimated. The theory of aggregates is not a logical theory; it draws, in Burge's version of it, upon our empirical and metaphysical beliefs concerning, e.g. what objects to count as individuals and how much and what kind of complexity is compatible with individualhood. It is not to be expected, therefore, that a logical theory will have the capacity to provide for every property and every aggregate *decision-procedure* for composition and division. That could be done, if at all, only if the theory of aggregates came to us completely determined. But is does not in fact. The indeterminacy in the theory of aggregates is no better illustrated than by the notion of a *mix*. Whereas aggregates are neutral with respect to such matters as the spatio-temporal configuration of their member-components, some complex objects that seem not to be individuals owe their identity very much to configurative and combinatorial factors. In the theory of aggregates, lemonade is such an entity. Lemonade is essentially a proportional entity. It is a mixture, a temporal phase of an aggregate, a phase during which "the relevant molecules are integrated within certain proportions". Moreover, one is dealing with the *same* lemonade "as long as one is dealing with the same aggregate and its components do not violate the appropriate mixture conditions". Very well, then, is an

Composition and Division 109

automobile, for example, a (complex) individual, a mixture or integrated phase of an aggregate, or just an aggregate? The question is a relatively open one, and it is not going to be closed by logic.

It will be agreed, then, that it is not for logic to complete the theory of aggregates, and that decision-procedures for composition and division are, for now at any rate, too much to hope for. Nevertheless, certain divisional and compositional generalities may be proclaimed with some hope of their being true. Here is just one example:

(a) If the term F is a mix-sortal, or a mix-mass term then F never divides over F-aggregate phases. *Example*: If 'is a car' is a mix-sortal 'is a car' does not divide over cars. If 'is lemonade' is a mix-mass term, 'is lemonade' does not divide over lemonade. Not all car-parts are cars; not all lemonade components are lemonade.

The fact is that for the great majority of properties, F, and aggregates, A, the composibility or divisibility of F with respect to A cannot be determined abstractly; one needs (i) an interpretation of 'F' and 'A'; (ii) a knowledge of the world; and (iii) a tolerable nose for metaphysical taxonomy.

We now introduce two definitions involving the notion of aggregate. Let $\{a_1, a_2, ..., a_j\}$ be a set of components of an aggregate, A, and let F be a predicate.

Def. Comp. H.: F is *compositionally hereditary* with respect to A if, and only if, every a_i has F then A has F.

Def. Div. H.: F is *divisionally hereditary* with respect to A if, and only if, A has F then every a_i has F.

For short, we may say that F "composes" or "divides" in A, or alternatively that F has the "compositional property" or the "divisional property" with respect to A.

Reflection will show that for the vast bulk of aggregates, some properties compose (or divide) in them and others do not. And in fact it is easy to show that the following pair of truths obtain.

(T1) *For any aggregate there is some property that does not compose in it.*
(T2) *For any aggregate there is some property that does not divide in it.*

For (T1), consider any property that applies only to a non-aggregate, such as the property of being an individual. For (T2), consider any property that applies only to an aggregate, such as not being an aggregate-component. Clearly these properties will not compose and divide respectively in any aggregate.

Clearly, too, an aggregate cannot contain itself as a component. Neither are aggregates components of other aggregates. Since it is impossible to have aggregates of aggregates, certain well-known problems of self-reference do not arise.

The following two principles will *not* obtain generally.

*(T3)   *If F divides in an aggregate, A, then F composes in A.*
*(T4)   *If F composes in an aggregate, A, then F divides in A.*

As a counter-instance to *(T4), consider the property of weighing more than two pounds. This property would compose where we are thinking of the aggregate made up by the parts of a machine. But it does not divide -that the machine weighs more than two pounds does not imply that every individual component will weigh more than two pounds. A similar counter-instance overturns *(T3).

We will now propose a framework for analysis of the fallacy via the triple {the concept of aggregate, *Def. Comp. H.*, *Def. Div. H.*}. It is by now clear that aggregates can be either linguistic or extra-linguistic, given the benign assumption that both linguistic and non-linguistic items may count as individuals. Thus our analysis can encompass both the mediaeval and modern concepts of the fallacy.

It is interesting that on the material rendering of "if-then", the set {T1, T2, *T3, *T4, *Def. Comp. H.*, *Def. Div. H.*} is inconsistent. Appendix B gives the proof. This, we say, clearly defeats the material construal of '→'.

5. ANALYSIS OF THE FALLACY

§9. We propose that there are two general forms of composition and division. The first kind of fallacy consists in arguing that because all the parts of an aggregate have F then the aggregate has F, or conversely. More specifically, the forms of composition and division are, respectively:

(F1) All the a_i have F.
    Therefore, A has F.

(F2) A has F.
    Therefore all the a_i have F.

So understood, the fallacies are simple converses of each other. (F1) and (F2) are logically incorrect because they do not hold for all A and all F. It would be erroneous (even fallacious) to infer however that (F1) and (F2) are never true, not true, that is, for some A and some F. Arguably, (F1) and (F2) may be true for some restricted subsets of A and F. This is made clear by a recent discussion in [164], [12], [39], that we will now briefly review.

The discussion of [164], taking Copi [40] as its starting point, is mainly concerned with inferences of the form below, where x is a whole and ϑ is a property.

(F3)   *All parts of x have ϑ.*
       Therefore, x has ϑ.

From the point of view adopted above, (F3) is simply one instance of (F1), but admittedly an important and interesting instance, even though Rowe's selection of (F3) for discussion is a reflection of the predominance of the modern view. Rowe argues that some inferences of the form (F3) are not fallacious, thus incidentally repudiating the thesis attributed to Copi [40] that all inferences of this kind are fallacious. Rowe considers a particular example of (F3),

(A5)   *All the parts of this chair are brown.*
       Therefore, this chair is brown.

Contrary to the validity of (A5), it may seem possible for a mysterious process of colour transformation to take place. Could it not be that when the chair parts are combined a transformation occurs such that the chair is a different colour? Rowe rejects this hypothesis, arguing that the hypothesis talks about the colour of the parts *before* they are connected together to form the chair [164, p. 88]. It is indeed possible that, before assembly, the chair parts could be a different colour from the assembled chair. Rowe argues that the question is rather: can all parts of *this* chair (not some pre-assembled counterpart) be brown where the chair is not brown? And Rowe asserts the thesis that this is logically impossible, i.e. (A5) is valid, although not first-order valid. Rowe concludes that sometimes arguments of the form (F3) are valid, sometimes not, but there is no "formal or general" way of sorting the valid from the invalid cases [164, p. 92].

Bar-Hillel [12] agrees with Rowe that (A5) is valid, as with this inference: If all the parts of x are made of pure iron, then x is made of pure iron [12, p. 125]. Contrary to Rowe, Bar-Hillel does think that there exists a general characterization of the fallacy of composition, although (in Carnap's terminology), he suggests that (A5) is not L-valid but requires "meaning postulates". Unfortunately, however, he does not inform us what that characterization is. Cole [39] is not much more help. He suggests that what is required to establish the validity of (A5) is an additional premiss,

> (P) When all the parts of a chair are a certain colour, the chair is that colour.

But as Hamblin [80, p. 21], notes, this is too easy a move. What we want to know is: in virtue of what general principles is (P) supposed to be a necessary truth, and how does this help us to rule on (F3)?

In short, the best we can say is that (F1) and (F2) are valid for some F (assuming Rowe's thesis is correct), but invalid for some F also. In other words, some F's are compositionally hereditary, some F's are divisionally hereditary, some F's are not compositionally hereditary, and some F's are not divisionally hereditary. Since the vast bulk of expressions that can go in for F are expected to be such that their logical behaviour is governed by principles outside first-order logic, Bar-Hillel is correct to conjecture that extra-logical "meaning postulates" are required. In Carnap's terms, these principles will be A-valid rather than L-valid. Perhaps this is a deeper sense in which composition and division may be said to be fallacies within language. Even in a logically perfect language (without A-postulates) these fallacies would not be ruled invalid. Still, it seems to us that the pursuit of a decision-procedure for these matters lies beyond the scope of abstract logical theory, and that this is well illustrated by the discussions just reviewed.

§10. The second general form of the fallacies of composition and division arises through confusion among the kinds of aggregates. The classical case of this form of the fallacy is the conflation of distributive and collective predication, well-known, as we saw, in the mediaeval logical literature. An example is the confusion between different kinds of collectivity. If A is a set-collectivity and its elements, a, b, and c, are individuals, then if we were to argue that A has F simply because each of a, b, and c individually (distributively) has F, then we would have committed the second fallacy of composition. Yet if A were an aggregate the self-same inference could well be sound.

Here the fallacy is not simply an invalid instance of (F1), as in the first kind, but a confusion between a valid instance of (F1) with an instance that may not be valid. In other words, the second varieties of the fallacy of composition (or division) trade on the fact that there are many distinct kinds of collectivities or wholes.[4] Thus for some F, the a_i may be compositionally (or divisionally) hereditary for some a but not for other a. So we see that a, as well as F, is a parameter that must be fixed to sort the valid and invalid instances of (F1) and (F2).

6. THE CAUSE OF THE FALLACY

§11. We have argued elsewhere, especially in [233], [225], and [234], that a fallacy is required to be (i) an incorrect (invalid) argument, and (ii) an argument that often *seems* to be valid. The study of fallacies has, broadly speaking, both a *logical* and *psychological* component, and both these factors must be considered in order to attain more than a surface understanding of a fallacy. Informal logic is thus partly psychological in its orientation; in particular, it must be geared to the dynamical texture of real-life argumentation. This does not mean that a fallacy is always first-order invalid. Nor does it mean that a fallacy must seem invalid to everyone, or everyone who advances it, or everyone who is a victim of it. Rather a fallacy is a systematic snare or delusion that is *generally conductive* to entrapment or deceit in argument. This dual aspect of sophismata was clearly recognized by William of Sherwood who distinguished in [210] between the *non-existentia* (non-existence, incorrectness) and the *apparentia* (semblance, seeming-correctness) of a fallacy. No analysis of a fallacy is complete without addressing itself to the question of its cause or occasion, the psychological question of the *modus operandi* of the fallacy in inferential belief-modification.

The basic explanation of the first variety of composition and division is that any composite object of perception or thought can be thought of as a Gestalt or as an aggregation of components, considered individually. As Hume might have said, the mind can slip easily from one mode of perception to the other. We can focus on the whole or pinpoint more discrete foci of attention. Thus the inferences from the one to the other domain are easily elided. The second kind of fallacy arises in particular because objects can be thought of as collectivities of many kinds. There seems to be no obvious limit to the number of distinct collection-forming operations that can be imposed on objects. And it is therefore possible and easy to confuse the kinds of relations that may obtain among them. All this seems rather obvious, and could no doubt be better explained by a Gestalt psychologist in more precise terms, but it is worth briefly remarking here how such effects can show themselves in certain truth-functional, quantificational and modal contexts.

There is a certain telescoping effect in English and other natural languages that sentential connectives can produce. The sentences

(1) John went to the store.

and

(2) Fred went to the store.

can be combined economically into

(3) John and Fred went to the store.

Yet there is more involved in (3) than the bare conjunction of (1) and (2). It is implied or suggested by (3) that John and Fred went *together*. Perhaps it is better to say that (3) is ambiguous or context-sensitive, sometimes having the implication of togetherness and sometimes not. This telescoping feature is not available in the propositional calculus - although English is systematically conductive to composition and division (of either variety) here, the standard formalization provides no mechanism for mapping the telescopic effect.

If we look to predicate structures, a possibility suggests itself. We could allow for conjunctions of individuals as in the expression '$F_{a\&b}$'. But it would be hard to know how best to interpret such expressions, and we shall not try to do so here. English is ambiguous in these regions, and whenever a sentence of the form 'a and b has F' is uttered, it can be interpreted in either of two ways.

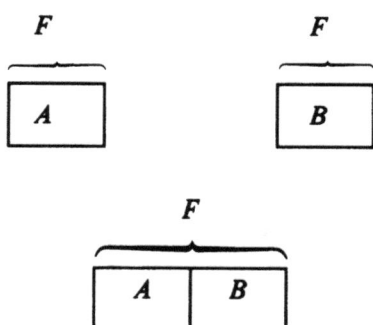

The telescopic capacity systematically allows for this ambiguity. When a and b are physical objects, for example, the expression 'a and b' usually suggests the spatio-temporal juxtaposition of a and

b, and the second interpretation is suggested.[5] Or to take a familiar sentential example, 'p and q are possible' can mean '$M(p \& q)$' or '$Mp \& Mq$'. Thus the natural language aids and abets the fallacy systematically. The word 'and,' merely one indicator of aggregation, is also a particularly nasty contributor to the second variety of the fallacy, since 'and' stands in for so many varieties of aggregative operations over and above mere conjunction.

Whately [208, p. 216], compares the psychological background of composition and division with the effect produced by the *Thaumatrope*, a card that has an object painted on each side - say a bird and a cage - where by quick rotary motion the illusion is produced of the-bird-in-the-cage. Similarly two distinct objects, if presented juxtaposed, again and again in quick succession, become associated together in the thoughts of an observer, even if they may be incompatible in reality. Through repeated transition we can be deluded into the notion of the actual combination of two things. It is this sort of phenomenon that seems to supply the *modus operandi* of composition.

7. IMPORTANCE OF THESE FALLACIES

§12. In modern times, perhaps because of the gerrymandered history of composition and division, it is often felt that these fallacies are of little real import in the actual analysis of argumentation, that they have found their way into textbooks simply as ancient Aristotelian relies. But the writings of Whately must convince one that the facts are otherwise. For one thing, Whately showed that three very important inductive-statistical fallacies are likely to be instances of composition.

(1) *The Fallacy of Overrating*: Whately [208, p. 214f.], sets out the following form of (invalid) argument.

> (F4) *If it is more likely than not, that these premisses are true: (i.e. that they are both true) it is more likely than not, that the conclusion is true: but it is more likely than not that the premisses are true: (i.e. that each of them is so) therefore it is more likely than not that the conclusion is true.*

In [221], we employed the relation of *befortification*:

p befortifies $q = Pr\ (q, p) > 0.5$

That is, p is said to *befortify* q where the conditional probability, q given p, is greater than one half. We could now represent the form of Whately's argument as follows, for two independent premisses, P_1 and P_2. A premiss, P_i, is said to be *befortified* (simpliciter) if it has a probability value of greater than 0.5.

(F5)   'P_1 & P_2' *is befortified* $\supset$ C *is befortified*
        P_1 *is befortified* & P_2 *is befortified*
        Therefore, C is befortified.

Here the illicit transition between the antecedent of the first premiss and the entire second premiss, is manifest. But in plain Latin, Whately's argument (F4) has considerable *apparentia*. The invalidity of (F5) is clearly shown by the failure of the following schema in the probability calculus (see the axioms in, for example, Skyrms [177]).

(F6)   *Pr (p & q) > 0.5 $\supset$ Pr(r) > 0.5*
        Pr(p) > 0.5 & Pr(q) > 0.5
        Therefore, Pr(r) > 0.5

Here then is a rather important class of applications of the fallacy of composition.

(2) *The Lottery Fallacy:* Whately's outline, [208, p.215], runs as follows.

(F7)   *The gaining of a high prize is no uncommon occurrence; and what is no uncommon occurrence may reasonably be expected.*

As Whately points out, the conclusion must be understood as "reasonably expected *by a certain individual*", whereas the middle term is most plausibly interpreted to mean "no uncommon occurrence to some one *particular* person". Thus here we have the fallacy of composition.

(3) *The Gambler's Fallacy:* Although Whately did not himself suggest it, an obvious extension of (1) and (2) is to the gambler's fallacy as an instance of division. Perhaps the gambler confuses the individual probability of p_1 & p_2 ... & p_j (where each is equal to 0.5, but where the probability of the entire sequence is decreased as each p_i is added) with the conjunctive probability. Thus the probability of some individual p_i is confused with the probability of a whole sequence of p_i's. Perhaps this is not the

complete explanation of all that is involved in the gambler's fallacy, but it does seem clear that the fallacy of division furnishes a significant part of the explanation of this important fallacy.

As is well-known, the fallacies of composition and division play an important role in certain modal arguments, and most notably in certain fatalistic or deterministic arguments, e.g. see Hughes and Cresswell [105, p. 27]. Again Whately [208, p. 215]: "He who necessarily goes or stays (i.e. in reality, 'who *necessarily* goes, or who *necessarily* stays') is not a free agent; you must necessarily go or stay (i.e. 'you must necessarily take the alternative'), therefore you are not a free agent".

Enough has perhaps now been said to establish that, from a point of view of significant errors and pitfalls of actual argumentation, composition and division are indeed fallacies of some genuine importance - easy enough to commit and mischievous enough to avoid committing.[6]

8. CONCLUDING REMARKS

We are in sympathy with Professor Bar-Hillel's refreshing remark in [12, p. 125]: "It is high time that modern logic textbooks should either leave fallacies alone or at least forget traditional classifications and start completely afresh". Continued acquiescence in the stale Standard Treatment perpetrates more than it explains the fallacies; and obviously some intelligent re-organization need to be done if the fallacies are to have a legitimate place in the study and teaching of logic. With respect to composition and division, we seem to have a choice: (i) we can continue to think of it mainly as an incorrect inference from part to whole or an equivocation on collective and distributive uses of words, taking our cue from the actual practice of the majority of modern texts; or (ii) we can take a much broader view of them, as we have suggested here, and think of them as two kinds of fallacy that pertain to such items as aggregates and collectivities.[7] Since aggregates can be both linguistic and extralinguistic, this is a perspective which covers both the modern and the mediaeval conceptions of the fallacy. Perhaps an even fitter choice is to compromise, and to think of the broad basis of the fallacy along the lines of (ii), yet bearing in mind that the range of cases most important to the teaching of students of the fallacies in elementary logic courses resides primarily in (i). Thus our efforts of analysis might be specially directed to the range of cases under (i) even while we recognize that the foundations of these fallacies extend back into (ii). This compromise, then, is what we will suggest as the most constructive basis for the further analysis of composition and division.

Appendix A

Transitivity of a.

To prove: $xay \land yaz \rightrightarrows xaz$

Suppose:
1. xay — Assumption
2. yaz — Assumption
3. $yaz \to Iy$ — (vi)
4. Iy
5. $(\forall z)(zay \to z = y)$ — From 4 by (vii), Simplification and Detachment.
6. $xay \to x = y$ — 5, Universal Instantiation.
7. $x = y$ — 1, 6, Detachment
8. xaz — 2, 7, Substitutivity of indenticals

Appendix B

Inconsistency of {T1, T2, T3, T4, *Def. Comp. H., Def. Div. H.*}.

Letting A, A', etc. be variables for aggregates,
    a, a', etc. be variables for components,
    F, F', etc. be variables for properties,
then a, a_1, ... are components of A, a', a_1', ..., components of A', etc.

Def. Comp. H. (F composes in A): $(\forall a) Fa \to FA$ [With '$\to$' for the material conditional.]

Def. Div. H. (F divides in A): $FA \to (\forall a) Fa$

T1 $(\forall A)(\exists F)[(\forall a)Fa \land \neg Fa]$
T2 $(\forall A)(\exists F)[FA \land \neg(\forall a) Fa]$

For a given A, let F be the property which satisfies T1:
    $(\forall a) Fa \land \neg FA$

F satisfies *Def. Div. H.*, for
    $\neg FA \to (FA \to (\forall a) Fa)$

T3 says: if F divides in A, F composes in A; so by *Def. Comp. H.*
    $(\forall a) Fa \to FA$

But by (1), Simplification
    $(\forall a) Fa$
    so FA — Detachment
    and $\neg FA$ — (1), Simplification

Similarly for T4.

Chapter 9
Post Hoc, Ergo Propter Hoc

1. THE PROBLEM

It is strange that the informal fallacies should strike us as such obvious breaches of thinking and advocacy, yet should have met with such little success in finding a respectable home within mature logical theory [80], [233], [221]. It might seem that respectable and mature logical theory is most mature and most respectable in the theory of propositions, and that its maturity and respectability in the other logical domains rapidly diminish in inverse proportion to the susceptibility of those domains to be reduced to the logic of propositions. But we are not anxious to promote so severe a view of theoretical accomplishment, and we shall suppose that, at the very least, the informal fallacies have a degree of systematicity that will at once advance our understanding of them nicely beyond the level of intuitive impressions, and also place into retirement the hopelessly inadequate accounts that litter too many otherwise admirable textbooks.

In the case of *false cause*, however, the theoretical orphanhood seems easily explained. We have heretofore lacked a causal logic [180], [31], [30], [76]. Common experience attests that the fallacy of *post hoc* is one of the most insidious and pervasive systematic misdemeanors of everyday argumentation: one need only reflect upon the conceptual mess attending such practically brutal realities as the causes of inflation or the causes of cancer. In fact, the special problem posed is this: is it possible to say something theoretically constructive about *false cause* without running afoul of general problems in the analysis of the causal relation? With a circumspection appropriate to the task, we here offer some thoughts about the structure of the *post hoc*.

2. BASIC CHARACTERISTICS OF CAUSATION

We do not presume to offer a wholly uncontroversial or complete explication of the causal relation, and we do not say that a complete and conclusive causal theory is a necessary component of an improved understanding of the causal fallacies. Nonetheless, a general characterization of some basic aspects of causation will be of some benefit to our project. It is, of course, always well to keep in mind the distinction between causation and the functions, correlations, equations, and graphs that constitute a part of the *evidence* for causal attributions. We also wish to emphasize the richly practical milieu in which the concept of causation is most naturally met. Talk of causation is an utterly familiar, accepted, and useful part of the language of the applied sciences in such fields as medicine, engineering, political science, economics, and psychiatry, all the more so where our interest is in the production or prevention of effects. Accordingly, we shall construe causal statements as arrayed against an assumed background of empirical and theoretical information that is not always statistically articulable. Statisticians often speak of this element in such terms as "common sense". So regarded, attributions of causality are bound by an understanding of *ceteris paribus*, in which reference is made to background information which is often empirical, yet not always explicitly statistical, and yet which plays an essential role.

To accommodate the factor of background information in causal explanation, John Anderson [4] introduced the concept of a field. This notion is given set-theoretical development by Suppes via explicit introduction of random variables [70, p. 110 ff.]. Suppes suggests that what is designated as a field is best seen as relative to the conceptual framework under discussion. Thus, says Suppes, "In one theoretical approach to the causal analysis of phenomena, the field will include only the consideration of macroscopic bodies and their characteristics, but in another, it will go deeper and consider as well atomic objects and their properties" [180, p. 75]. When a given framework of variables is extended by consideration of additional variables, what may have been a genuine causal relationship relative to the first field may turn out to be causally spurious relative to the second, "richer" field. A number of supporting examples can be drawn from learning theory, and they are analyzed in detail by Suppes. So relativized, causal explanation is naturally and usefully regarded as resting not upon laws *tout court*, but upon quasi-laws [163], [30, 7.3.3]. Quasi-laws are generalizations with tacit qualifications which cannot be fully articulated or spelled out in a finite, manageable set of explicit sentences.[1]

Let us, in any event, set out, schematically and briefly, various analytic perspectives on the causal relations that have attained some degree of importance in the literature. Our purpose in so doing is to seek for an intuitively satisfying core understanding of causation. It would be priggish to ask for more.

The *Humean approach* is paramount among causally non-familial analyses,[2] that is, among analyses set forth in an idiom free of causal locutions. From Hume's remarks[3] it is customary to extract this schema:

x caused y if, and only if,
H(1) x and y occurred in some same close neighbourhood
H(2) y succeeded x
H(3) occurrences of the x-kind in given close neighbourhoods are succeeded by occurrences of the y-kind in those same neighbourhoods.

An historically influential perspective, the Humean position, for all its problems, thrives in the contemporary literature. Suppes' theory, for example, briefly mentioned above, is broadly Humean in its orientation.

The dense and powerful litany of objections to the Humean view extends from Reid, to Ducasse, to Madden [157], [53], [132]. Numerous writers have remarked that Hume, unwittingly perhaps, ruled out simultaneous causation. Reid seized on the crucial importance of controlling the distinction between accidental and genuinely causal correlations, and Ducasse, many generations later, repeated the objection with even greater force. Then, too, the Humean definition runs into the heavy weather of talk of "kinds" of events, and of the possibility of unique happenings, and so on. It would appear that there is little chance of rescuing the Humean account as a satisfactory definition of "x caused y", but there is no doubting its value as a rich source of basic intuitions.

Some contemporary views, typically sympathetic to the Humean account, take the form:

x caused y if, and only if,
L(1) y did not precede x
L(2) a statement of the occurrence of y is deducible from the laws of nature together with a description of initial conditions including the occurrence of x
L(3) yet a statement of y's occurrence is not deducible either from x's occurrence alone, or the initial conditions alone or from the laws of nature alone.

With this, the *laws of nature account*, problems also rage. For all the intuitive advantage of tying causation to lawful correlations bound by certain restrictions, not all laws are causal (e.g., the Boyle-Charles law). Moreover, on this view, too, if one is not careful, simultaneous causation is ruled out. Also, if we are not careful (in particular, if we model laws after the generalizations of such abstract systems as classical mechanics), we risk committing causation to system-closure and thus risk suppressing the inarticulable background aspects of the causal idiom and the related susceptibility of causality to manifest itself in quasi-laws. Not least among the difficulties of the laws of nature account is that a philosophically satisfactory specification of "law of nature" is a formidable undertaking indeed. Even a perfunctory review of the contemporary literature reveals the prominence of the view that a universally quantified generalization is a law if its truth, if true, is preserved under subjunctive re-expression. Yet, even so, the subjunctive conditional itself stoutly resists the blandishments of philosophical analysis. We will not here trouble the reader with the lively controversies at mid-century and fifteen years, and more, beyond, with the squalls and tornadoes gusting over the familiar landmarks of Chisholm [35] and Goodman [77]. It suffices to remark that even the relative successes of the later modal analyses of Stalnaker, Lewis, Sobel, Woods, and others, do not leave the laws of nature approach trouble-free. For one thing, a Stalnaker sort of analysis [178] commits one to the proposition that, for every A and B, A either subjunctively implies B or subjunctively implies not-B. And in any such modal, or possible worlds approach to subjunctive conditionality, additional difficulties lie in wait. For example, in a Stalnaker sort of theory, A subjunctively implies B in this, the real world W, only if there exists a unique world W* maximally similar to W except that in W* A is true (and so too is whatever follows from A), *and* B is true in W*. Some writers (Lewis, for example) have queried the justification of supposing W* to be unique. But there is another problem as well. It has to do with the theoretical motivation of *world-similarity*. Why, we should ask ourselves, is it that the career of the consequent B in similar circumstances should determine the career of "If A were the case, so would B be the case" in present circumstances? What burdens does the similarity relation bear? Very crucial ones, as is evident from the following. Suppose that W*, the world alternative to our world W, were a world which offered its hospitality to generalized accidents but were utterly hostile to causal correlations. In *that* world the semantic career of B would be irrelevant to the domestic semantic career of "If A were the case, so would B be the case". No, W* must not just be an alternative to W and must not be similar in certain respects; it must be radically similar, differing only in ways having to do with the truth there of the antecedent, A. Thus the theoretical force of "is a world similar, in the theoretically relevant respects, to W" is that the alternative universe not be hostile to causal correlations, that is, that W* be a causally lawful world.

So, one is here at the brink of circularity. A law of nature backs its corresponding subjunctive conditional; a subjunctive conditional is true (in this world) whose consequent is true in a world in which the antecedent is true *and* which differs from our world in no other way (except for what follows from the consequent's truth). Yet such a world is a world of natural laws.

Still, the intuition of causal lawfulness is powerful. Perhaps a better analytic expression of it lies elsewhere, in what some writers call the *necessity in nature approach*: [60], [22], [65].

x caused y if, and only if,
N(1) y did not precede x
N(2) x is a complex of events each component x_i of which is logically necessary for y and $x_1 \cup x_2 \cup \ldots x_n = x$ is logically sufficient for y.

But here, too, the problem of simultaneous causation recurs. What is more, there is room for thinking, even if causation does amount to necessary correlation in nature, that logical entailment over-expresses the idea; entailment is just a shade *too* necessary for causal comfort. Better, then, that we seek a more moderate expression of causal necessity. Burks, indeed, is the classical source. In his causal logic, to which we return below, Burks develops a *modal approach*, a principal difference of which from the previous approaches is that it is a causally familial analysis. The primitive notion in Burk's analysis is the modality causal necessity, already an idiom of the causal family. It arises by definition that x causes y if and only if the material implication of x by y is causally necessary, where x and y are sentence variables. A principal point about the logic of causal necessity is precisely that it is weaker than logical necessity. In some versions of it, [31] Burks gives a partial expression of this feature by means of the axiom:

$$\Box \emptyset \supset \boxdot \emptyset$$

of which the converse does not obtain. This quickly gives rise to trouble, however, putting "p ⊃ p" for ∅, we have it that "☐(p ⊃ p)" obtains and hence, by this axiom, that " ⊡(p ⊃ p)" also obtains, and so that p causally implies p. Attributions *causa sui* are perhaps a trifle too Spinozistic for contemporary ears to compete for serious attention. If, in the natural necessity approach, causality is always a matter of logical necessity, then in modal approaches logical necessity is always causal necessity - kinships, in each case, not to be looked upon with favor.

How, then, does one characterize this "pushiness" of causation? One writer, Von Wright [189], among others [71], addresses the matter by means of the following *anthropomorphic account*. On this view, causation is paradigmatically a matter of someone making something happen. In one version characteristic of this position,

x caused y if and only if
A(1) y did not precede x
A(2)   someone, P, brought it about that x occurred, whereupon y ensued.

Of course, causal attributions are not in fact confined exclusively to the domain of human intervention. It is necessary then that A(2) give way to a more appropriately expansive counterpart. One reasonable-seeming prospect might be:

A(2')   Someone, P, had he brought it about that x occurred, thereupon y would have occurred.

Wherewith the anthropomorphic account defers, to a significant degree, to a subjunctivity account. Still, for all the unsettled state of their frontiers, certain causal intuitions remain relatively unscathed.

Two states of affairs, then, could be said to stand to one another in the causal relation where; 1) the one state, the cause, does not temporally succeed the effect, [30, 7.4.1], [130]; 2) cause and effect are reasonably proximate in space or time or both, (for example, often causation is triangulated with some physical process or relation); 3) taken together with a set of other (often unspecified or roughly specified) necessary conditions (none of which is individually sufficient) the cause constitutes a sufficient condition of the occurrence of the effect for appropriately qualified senses of "necessity" and "sufficiency"[4]; 4) the cause is isolable from the other necessary conditions in the sufficient set by pragmatic criteria having to do with ease of practical manipulation (this involves the question of "pragmatic relevance").[5]

An extensive new analysis of the causal relation is to be found in recent work by Burks [31], [30]. This analysis is given in two parts. First, the formal properties of causal necessity are given by a modal logic of a statement-operator. This logic is presented as an algorithmic structure and applied as a model of ordinary and scientific causal discourse. Syntactic structure for the causal modalities is given by Burks through decision algorithms, and an abstract interpretation for the language is yielded by simple modal models. Secondly, a concrete interpretation of the language is given by treating the logic of causal statements as a model of natural language. Non-probabilistic theoretical laws of nature are treated as causally necessary. Some important features of causal laws are modelled by means of a relation called by Burks "non-paradoxical causal implication" (*npc*), defined in terms of causal implication as follows: φ *non-paradoxically causally implies* ψ if and only if φ *causally implies* ψ, and φ is *logically independent* of ψ, and φ is *causally possible*, and ψ is *not causally necessary*. This definition drives a wedge between the causal and logical modalities (1) by requiring that the antecedent and consequent of a causal law be logically independent, and (2) by excluding the paradoxes of logical implication and the corresponding paradoxes of causal implication [30, 6.6 *vide* (14a) through (15f)].

Next, Burks defines a relation of *elliptical causal implication* (*ec*) in terms of *npc*, as follows:

ψ elliptically causally implies ϑ if and only if there is a true statement φ that causally implies ϑ when conjoined to ψ.

Readers can now begin to see how Burks' relation of elliptical causal implication adumbrates the intuitive notion of causation sketched above. Burks sets out some basic theorems and properties of these causal relations, and causal notions are extended into inductive contexts in considerable detail, but we shall not outline these matters further here.

In general, we shall follow Burks' analysis of the causal relation where exact specification of the causal idiom is called for, although most of what we have to say at this stage may informally be couched in the intuitive notion of "cause" set out above. We intend our analysis of the *post hoc* to coincide with Burks' treatment of causation, however. In fact, we look upon our excursus into the *post hoc* as a potentially interesting application of Burks' causal logic.[6]

This is not to imply that Burks' system is trouble-free, or the only source of sound analytical counsel. Recent attempts of a systematic type, helpfully chronicled by Domotor, [52] show that some interesting work is currently in progress, although results are rather austere and heterogeneous. Few of us are under any illusions about the difficulty of the general problem. One new approach, which we shall mention briefly because it seems to us to show special promise, takes as a basis a propositional calculus with classical negation and conjunction but utilizes an essentially relational disjunction and conditional. This is a system designed principally by R.L. Epstein and R.I. Goldblatt[7] (in a group project at the Victoria University of Wellington during 1975-76) to provide a more flexible approach to conditional expressions. The early motivation for this approach was some work of Douglas Walton on the logical form of action expressions. The system proceeds by adding to the semantics of classical PC a relation on propositional variables, $r(P_i, P_j)$. r is thought of as expressing a notion of *semantic relatedness*, and is symmetric and reflexive, but not (initially, at least) transitive. The "truth table" for $\rightarrow$, the relatedness conditional connective, is set up in such a way that $P_i \rightarrow P_j$ is classical if and only if P_i bears the relation r to P_j. The system closely resembles classical PC, retaining *modus ponens* and many of the usual theorems - certainly this is one reason why it is so interesting - but it notably lacks the rule of addition ($P_i \rightarrow (P_i \vee P_j)$), importation (($P_i \rightarrow (P_j \rightarrow P_k)) \rightarrow ((P_i \wedge P_j) \rightarrow P_k)$), and the De Morgan Equivalences. Disjunctive syllogism is, however, preserved. A set of axioms has been given, completeness shown for them, and the system has been extended to include quantifiers. An advantage of this approach is that it allows us to vary the conditions on r, thus yielding up a more flexible approach to conditionals, and so it provides a new basis for extending conditionals in the direction of causal contexts.

Now somewhat more precisely characterized, we can see how the causal relation gives us at least a tentative basis for proceeding, despite the causal skeptic who argues that causation is an intrinsically incoherent notion. Russell, for example, once argued that the concept of cause is intrinsically unclear and ought to be excluded from the language of science altogether [166]. Russell claimed that "causation" is a term belonging to our "metaphysical" heritage and that it expresses no scientifically useful concept. Scientists, in their more technical moments, at any rate, talk about correlations and functions, but this is not, perhaps, more true, if it is true at all, in the theoretical[8] than in the applied areas of science. In these latter areas, causal terms are a familiar and accepted part of the language. This is so because in the applied sciences there are inarticulable elements of background information that are necessary to account for in making rational decisions and predictions. It is the presence of such inarticulable bodies of information in applied scientific explanations that makes for the particular utility of the concept of causation. We shall see this factor reflected in the causal fallacies time and time again. It is specifically recognized in adequacy condition (3) above.

It would seem, in fact, that Russell, conflating the notion of cause with notions such as function, has himself committed a kind of fallacy in blurring the distinction between evidence for causation and causation itself. The urge to commit this fallacy is perfectly understandable in severely abstract theoretical contexts, where causation is not regarded as an essentially practical concept. But once we adopt the latter point of view, the common distinction made by statisticians between cause and function defies conflation.

> The mathematician is interested in functional relationships for their own sake. But he is also interested in them practically since very often these functional relationships form a good enough approximation to practical cases. ... The thing to be clear about ... is that a functional relationship in mathematics means an exact and predictable relationship, with no ifs or buts about it [140, p.275].

Causal relationships, on the contrary, always have "ifs" and "buts" attached. Herein lies their special value. Thus Russell's point, more soberly interpreted, would seem to be this: the more abstractly mathematical one's scientific pursuits, the less easy it is to find obvious among the concepts there dealt with any close reconstruction of everyday causation. But this is a far cry from saying that modern physics has expunged causation from its repertoire of relations.

3. STRUCTURE OF POST HOC ARGUMENTS

One might be tempted to conjecture that the fallacy of *post hoc* has the following form:

(A8)   There is a positive correlation between φ and ψ.
Therefore φ causes ψ.[9]

The difficulty, of course, is that this form of argument may, so far as it goes, be inductively *correct*, for even a single instance of positive correlation may, in the total absence of contravening information, be *some* (not to say very impressive) evidence of causation. In a way, this factor is the albatross of the formal analysis of the *post hoc*. *Ceteris paribus*, a correlation is sometimes perfectly good, if minimal, evidence of the existence of a causal connection.[10] However, we think that a more perspicuous characterization of the fallacy involves the spurious inflation of the evidential value of the correlation owing to the suppressing or failing to take into account other causally relevant factors of various kinds. To the extent that such causally relevant factors, of the kinds we shall examine, can tentatively be characterized under standardized headings, we suggest that any correct inductive argument from correlation to causation, requires five premise-types, as represented by the following scheme:

(P1) There is a positive correlation between φ and ϑ.
(P2) It is not the case that ϑ causes φ.
(P3) It is not the case that there is a third factor, ψ, that causes both φ and ϑ where φ does not cause ψ.
(P4) There are no relevant instances of φ-and-not-ϑ.
(P5) φ is pragmatically relevant.
Therefore, φ causes ϑ.

Here, then, is a total of five independent kinds of conditions, represented by (P1) to (P5), against which a claim of causation should be tested.[11] This suggests that we characterize the fallacy of *post hoc* as involving the spurious argument from (P1) to the conclusion by suppressing one or more of (P2) to (P5). Accordingly, we may distinguish four gross varieties of the *post hoc*, depending on which of (P2), (P3), (P4), or (P5) is suppressed. We develop this perspective in Section IV.

Other ways of expressing something interestingly close to the structure of the *post hoc* also suggest themselves. For example, Reichenbach ordered events on a world line ("genidentity chain" or "causal chain") by a relation of *approximate spatiotemporal coincidence*.[12] Accordingly, one might make use of some such relation to characterize inferences of the following form:

(B8) φ is approximately spatiotemporally coincident with ϑ. Therefore, φ causes ϑ.

Now such an inference would be invalid within Reichenbach's theory, since other requirements, such as the *local comparability of time order* and *possible connectedness* are needed as additional premises.

Alternatively, the latter notion of connectibility is defined by Berger [20] as a binary symmetric and reflexive relation K of *causal connectibility* defined on a four-dimensional differentiable manifold. Ensuing from this could be yet another plausible model of *post hoc* reasoning.

(C8) φ is causally connectible with ϑ. Therefore, φ causes ϑ.

For, as Berger points out, [20, p. 56] the claim is that K (φ, ϑ) conveys no information about the point (φ or ϑ) at which the later (or earlier) event must occur for φ and ϑ to be *actually* causally related. So understood, the logical blunder would be a modal fallacy of arguing from possibility to actuality.

Space does not permit an attempt here to mesh the frameworks suggested above with that of the previous paragraph. Rather, we leave this for the time being as an open problem in the theory of the *post hoc*.

One qualification: we shall, by and large, limit ourselves for ease of exposition to cases where causes or effects are thought of as individual dateable events, φ, ψ, and ϑ. It is well known that causal language is ambiguous in its capacity to range over "generic events" as the relata of the causal relation. It is often said, for example, that smoking causes cancer (generally) in a manner indicating that no dateable, particular instance of smoking or cancer is exclusively being referred to. Our treatment will not be specifically addressed to these cases; although it is to be hoped that our analysis might usefully be extended in this direction after the method of Burks, it is not a project we attempt here. Our preoccupation will be with singular causal statements rather than causal laws.

Norman Swartz has pointed out to us that the scheme represented by (P1) to (P5) is more accurately seen as two different causal paradigms, one of which ought to be stated in terms of probability values and the others in the language of correlation coefficients. Readers uninterested in the niceties of form might pass on to Section IV immediately, but for those who require a more exact approach to the analysis, we here indicate briefly how our scheme should really be thought of as *two* schemes, or at least two variants of one superscheme. The first case is that where two variables vary together over a range of values ((P1) is supposed to capture this case), and second, the case where the two variables can take on only binary values ((P2) would cater to this case). An instance of the first would be plumbing the relationship between the percentage of voter registration in a city and voter turn-out on election day; each variable can take on values between 0 and 1. An instance of the latter would be the ingestion of arsenic (values 0 *or* 1) and death (again values 0 *or* 1). Now there may be no cases in which ingestion of arsenic is not followed by death shortly thereafter; we suspect that there *is* causal relation here. But if we examine the correlation between the whole series of arsenic-takings over a large population we will get something like

"0, 0, 0, 0, 0, 1, 0, ... 0"; the corresponding values for death will be "1, 0, 0, 1, 0, 1, 0, ... 0". Here we will get a high correlation between the two series, just because there are so many zeroes in the first series and so many zeroes in the second (or, to put it another way, almost all persons who are alive today, will be alive tomorrow). The correlation, however, seems to be spurious in establishing the hypothesis that ingestion of arsenic causes death. In short, when we are looking at discrete events (binary values) we ought *not* to be invoking correlation coefficients. In the latter cases we ought to be looking at probability figures: what is the probability of the ingestion of arsenic being followed by death? Or, what is the relative frequency of death among the class of arsenic users? Since, for one thing, Probability (A, B) need not equal Probability (B, A) while Correlation (A, B) does equal Correlation (B, A), we need to distinguish between the two types of cases generally. And for discrete events, it seems that we shall want to put our scheme in terms of probability values rather than correlation coefficients. Thus, a more precise statement of the scheme represented by (P1)-(P5) would involve two parallel sets of condition depending on whether the variables are continuous or binary.

We now turn to a review of the practicalities of the *post hoc* as a real-life error of argumentation in order to provide substantiation and application of our schematic analysis. In the next section, we deal with some actual instances of argumentative deficiencies often associated with *post hoc* reasoning.

4. SEVEN POST HOC FALLACIES

It is possible to distinguish within traditional treatments of informal fallacies at least seven significantly distinct causal fallacies, some of which have sometimes been classed under such standard headings as *post hoc ergo propter hoc, non causa pro causa, secundum quid, false cause* and the like. We here present a quick sketch of the main thrust of each of these seven sophisms, in each case offering at least one concrete example. The object in each case is to plumb the example in question for the essential wrongfulness of the fallacy it illustrates.

1. *Concluding that ψ was caused by φ just because ψ temporally followed φ*. A stock example:

> (A1) I took a dose of Sinus Blast and a couple of days later my cold cleared right up.

The suggested conclusion, namely that taking Sinus Blast *caused* the cold to disappear, is fallacious: it may be counterclaimed that the cold may well have disappeared just when it did or perhaps even sooner if Sinus Blast had not been taken. A second specimen is exemplified in the behavior of a passenger on board the doomed Italian liner, *Andrea Doria*.

> (A2) On the fatal night of Doria's collision with the Swedish ship Grisholm, off Nantucket in 1956, the lady retired to her cabin and flicked the light switch. Suddenly there was a great crash, and grinding metal, and passengers and crew ran screaming through the passageways. The lady burst from her cabin and explained to the first person in sight that she must have set the ship's emergency brake [64, p. 166].

Quite clearly, attributions of a causal connection from a *single instance* of a given simple sequence of events greatly risk a fallacy. Single instances have a way of being coincidences.

2. *Concluding that ψ was caused by φ just because there was a positive correlation between some previous instances of φ and instances of ψ.* A stock example:

> (A3) Near perfect correlations exist between the death rate in Hyderbad, India, from 1911 to 1919, and variations in the membership of the International Association of Machinists during the same period [120, p. 368].

In the absence of further evidence of a causal connection between these remote sets of data, concluding that one is the cause of the other would be absurd. While in certain circumstances a high positive correlation may be good evidence of a causal link, in this case "common sense" informs us that it is extremely unlikely that the two sets of the phenomena are causally connected.

3. *Reversing cause and effect*

> (A4) The people of New Hebrides have observed, perfectly accurately, over the centuries, that people in good health have body lice and people not in good health do not. They conclude that lice makes a man healthy [104, p.98].

A little further observation reveals that, whereas lice were the norm among these people, when anyone took a fever and his body became too warm for comfortable habitation, the lice departed. Insofar as there is a causal relation between the lice and good health, the above conclusion reverses cause and effect. The causal relation is non-symmetrical; an additional inference would be required to establish which correlate might be the cause of the other. As Huff points out [104, p. 89], commonly there is a genuine causal relation where it is not evident which of a pair of correlates is the cause and which the effect, e.g., the correlation between income and ownership of stocks.

Social science methodologists have used the techniques of partial correlation analysis [174], [21] and path analysis [88], [244] in identifying errors of causal inference. The types of errors recognized by these methods tend to correspond to the categories of fallacies we are here discussing. In particular, a Type D error is defined by Deegan [49] as one which occurs when all the variables in a true model are utilized but where a true independent variable is hypothesized to be dependent, or conversely. The conditions under which this kind of error can arise are: 1) when the theory fails adequately to prescribe the temporal sequence of events or actions; 2) where an investigator fails to grasp the technical requirements of the inquiry; or 3) when the system is assumed to be closed and the analyst chooses as the "correct" model the causal sequence that explains the largest proportion of variance.

4. *Concluding that φ is the cause of ϑ when both are the effects of a third factor, ψ.*

A correlation between φ and ϑ may be indicative not of a causal relation between φ and ϑ,

$$\varphi \to \vartheta$$

but of some third factor that causes both φ and ϑ, thus accounting for the correlation,

$$\psi \swarrow \searrow \\ \varphi \quad \vartheta$$

The latter state of affairs is called a spurious correlation, and it contrasts with the deceptively similar situation where ψ is an *intervening variable* between φ and ϑ

$$\varphi \to \psi \to \vartheta$$

allowing a genuine causal relation to obtain between φ and ϑ.[13] In both cases we have

$$\psi \to \vartheta$$

The defining feature of spurious correlation is that φ does not cause ψ, although the converse obtains. Two examples will make the contrast clear. [245], [174]

> *Example 1*: In a survey on factory absenteeism it was found that married women had a higher rate of absenteeism than single women. Later it was found that the absenteeism rate among married women was almost as small as that of single women if both have little or no housework, and that absenteeism among single women was almost as great as that of married women if they too have a great deal of housework. A further survey showed that, as one would expect, married women generally had more housework. We could correctly conclude:

> getting married → more absenteeism.

But a more complete explanation may be found in the intervening factor of housework:

> getting married → more housework → more absenteeism.

Here we have a genuine causal relation mediated by a third factor.

> *Example 2*: it was found that married persons ate less candy than single persons. A second look at the data showed that if married and single persons of equal age are compared, the correlation between marital status and candy consumption disappears. Here it would be misleading and incorrect to conclude:

> getting married → eating less candy.

The correct conclusion is that getting older is the operative factor in both increased likelihood of marriage and decreased candy consumption. Thus:

$$\text{getting married} \swarrow^{\text{getting older}} \searrow \text{eating less candy.}$$

The critical distinction, between this case and the previous example is that here it is incorrect to conclude:

getting married → getting older.

If we could prevent people from getting married this would still not stop them from getting older. But conversely the causal relation may obtain because if we would stop people from getting older, this would presumably have the effect of reducing the incidence of marriage.

It is also interesting to note that even in the case of intermediate causation:

$$\varphi \rightarrow \psi \rightarrow \vartheta$$

a fallacy can arise through omission of or addition of intermediate factors and through consequent mischaracterization of the nature of the causal relation between φ and ϑ. Take the case of Mr. X who, while driving in traffic, repeatedly observes that whenever he applies the brakes, the defroster fan squeaks. He concludes that the brakes are connected to the fan. A more mechanically sophisticated observer might infer that application of the brakes causes deceleration of the vehicle and tilting of the fan mechanism, which in turn causes the squeak.

It is especially revealing to note, as Zeisel [245, p. 88] points out in his analysis of the candy and housework examples, that the final test of the merit of the distinction between the two cases is an essentially *practical* matter. If a factory manager were to institute the policy of discouraging female employees from marrying, this policy would effectively tend to reduce absenteeism. Remaining single means less housework and less housework means less absenteeism. But suppose a candy manufacturer were to succeed in discouraging large numbers of people from marrying. Would this lead to an increase of candy consumption? Clearly not. Keeping people younger would have the desired effect, but preventing them from getting married would not keep them from getting older. Thus, ultimately, the distinction between "true" and "spurious" causation rests on the practical question of whether φ is a practically necessary condition of ϑ. If I stop φ will ϑ stop too? Or can I only stop both φ and ϑ through ψ, a third factor?

The fallacy in question arises in confusing the two cases below.

(1)  (2)

intervening variable   *antecedent mutual cause*

The result of the confusion is a mistaken equation of the correct inference, (1*) with the incorrect inference on the right, (2*).

(1*)  (2*)

correct   *incorrect*

(1*) is correct, provided the causal relation is transitive. (2*) is incorrect: (2*) may, though fallacious, seem correct because the illicit assumption of symmetry of the causal relation,

$\psi \to \varphi$
Therefore, $\varphi \to \psi$

creates the suggestion of an equivalence with (1*). What distinguishes the fallacious (2*) from the correct (1*) is that $[\varphi \to \psi]$ fails in the former, but not in the latter.

5. *Confusing causation and resemblance.* David Hacket Fischer offers a vivid example of this fallacy [64, p. 177].

(A5)   The Picts constructed brochs and souterrains that where small, dark, and mysterious. Therefore the Picts themselves were small, dark, and mysterious.

The fallacy seems to form the basis of many traditional folk remedies, e.g., that a bloodstone will stop bleeding.

6. *Citing a pragmatically otiose necessary condition.*

Example:

(A6)   Smith drowned because he did not learn to swim when he was young.

Hence, the causal attribution strikes the ear as far-fetched. Generally one would be more interested in more immediate factors, more at hand to possible reversal and more proximately previous to the drowning. However, in a context of a coach's

138   Chapter 9

presentation of the benefits of swimming lessons, the causal claim above might not appear quite so remote. Again, causal "talk" is strongly linked to practical production and prevention. Thus, it can be true that φ is genuinely a necessary condition of ψ, yet, nonetheless, be a fallacy to conclude that φ caused ψ, where control of φ is impractical, irrelevant, or impossible.

7. *Overlooking or suppressing information that may run counter to the apparent trend of the correlation.*
Example:

> (A7)   It is easy to show by figures that the more it rains in an area, the taller and better the wheat grows. Conclusion: rain is good for the crops.

The problem here is that the positive correlation holds up to a point beyond which it takes on a negative significance. [104, p. 91] The fallacy is reminiscent of the familiar abuse of sampling theory in which an unrepresentative sample is selected. An instance of $\ulcorner \varphi \ \& \ \neg\psi \urcorner$ can overturn a causal allegation of $\ulcorner \varphi \rightarrow \psi \urcorner$; negative correlations must also be considered. Or, as Mill would have put it, the method of agreement needs to be supplemented by the method of difference.

5. COMPLETENESS

The structure of Section III can be integrated with the sophisms of Section IV in a way that plausibly displays a set of covering relationships between premise (P1) and sophism (1) and (2), premise (P2) and sophism (3), (P3) and (4), (P4) and (7), and (P5) and (6). Some discussion of those matchings is now in order. (We omit treatment of (5)). Although it is a kind of causal fallacy, we do not regard it as essential to *post hoc*, but rather as a kind of fallacy of resemblance or analogy.)

It is also interesting to note that in (1) and (2) we rejected φ as a cause of ϑ, because if φ had not occurred we believe that ϑ might (or would) have occurred anyway. A counterfactual analysis, after the fashion of Lewis, might therefore restructure (P4) to read: if φ were not to obtain then ϑ would also not obtain. Our analysis of the *post hoc* could be adapted to the counterfactual treatment of Lewis', much in the way it was adapted to Burks' analysis, but we do not pursue the point here.

Concerning the second premise, (P2) cannot be established without considering an independent correct inductive causal argument. In practice, statistics texts often establish (P2) by appealing to the temporal precedence of φ to ϑ, or by citing "common sense". It seems to us more realistic explicitly to use the term "cause" in (P2) (and thereby frankly acknowledge that the analysis may be circular, or only partially definable, perhaps by appeal to a recursive structure), rather than to try to state a noncausal (P2) that will probably turn out to be inadequate. As elsewhere, however, we wish to avoid causal dogmatism, so we do not exclude this possibility entirely, and commend the problem along with others we have suggested, as important for further research.

A consideration brought to our attention by David Loewen suggests that the set of premises (P1)-(P5), given the broad conception of causation we began with, may be *essentially incomplete*, for there may be a situation in which (P1)-(P5) obtain yet in which the causal relation is not warranted. Consider a case of what seems to be a perfectly fortuitous positive correlation such as that between the stork population and the (human) birth rate in a given area. Here (P2) to (P5) may be satisfied, yet we might nevertheless be quite reluctant to presume a causal connection between the storks and the babies.[14] It would be premature and ill-advised seriously to entertain any claims of completeness for our analysis of the broad, intuitive notion of causation met with in the opening sections of this paper. Even so, perhaps the case of the storks is defeated by the parameters of "pragmatic relevance" cited by (P5). However, since we are not in a position to give an analysis of these factors in any rigorous way, such claims or counterclaims of completeness are best held in abeyance. Both Lewis and Burks also set these apparently slippery questions aside in attempting to deal with the somewhat better-behaved core of causality represented roughly by (P1) to (P4), and we shall content ourselves with a kindred diffidence here.

6. CORRECT AND INCORRECT CAUSAL ARGUMENTS

We have attempted, in Sections III, IV, and V, to say something helpful about *post hoc* from a point of view of the applied logic of the informal fallacies, but without attempting a full-scale analysis of the causal relation intuitively adumbrated in Section II. Nevertheless, in describing the fallacy of *post hoc* as an invalid, incorrect, or deficient *argument* that deviates from the correct form postulated in Section III, our point is clearly parasitic on some theory of *correct causal argument*. We, therefore, conclude with some general remarks concerning the relation between our findings here and the larger project.

We are inclined to think of a causal argument after the fashion of Burks (*Cause, Chance, and Reason*, 10.3.2), by way of causal model of standard inductive logic constructed from causal cellular automata.[15] The form of a possible causal law is:

(L) For all cells and for all times, the occurrence of property φ in region N(c) at t causes the occurrence of property ψ in cell C at t'.

The causal universal (L) deductively implies a corresponding material universal,

(U) For all cells and for all times if φ occurs in region N(c) then ψ occurs in cell c_i at t + 1 (with "if then" here expressing material implication).

The material universal, in turn, deductively implies the instance-statement,

(I) If φ occurs in region $N(c_i)$ then ψ occurs in cell c_i at t + 1, for each cell-moment $<c_i, t_i>$.

Two argumentative relations that could obtain between I and L are as follows; If I is false then L is false, since L deductively implies I. This is the case of direct refutation. If we were to take up one elementary notion of confirmation, one by which A affords a measure of confirmation to B when the conditional probability, A given B, is greater than the prior probability of A alone then we can see that (P1)-(P5) bears such a relation to the causal conclusion of our scheme. Elsewhere,[16] we have tried to show how such a confirmation relation (we call it *befortification*) can be used to convey a notion of "correct argument" appropriate to the study of the kinds of arguments open to the informal fallacies. Remarks, here, are confined to indicating in only a programmatic way what sorts of theoretical resources might be brought to bear in deepening the analysis of the *post hoc*.

One final foundational comment: we are coming to think that a relatedness semantics has advantages over approaches such as Burks', based as they are on the classical material conditional for PC. Not only does the relatedness method offer greater flexibility, but it avoids the paradoxes in a more natural way. Moreover, as David Lewis has suggested,[17] the relation r may be thought of as an assignment to sentences of subject matters such that sentences are related if and only if their subject matter overlaps. We might then think of r as expressing "semantic relatedness". But we would also suggest that r could equally well be thought of as modelling the notion of "spatiotemporal coincidence" or even "causal connectibility", met with earlier, thus giving rise to a practical new application of relatedness semantics to the causal idiom.

We reject the thesis, often ventured in the social sciences, that causal arguments are fallacious as such. We suggest instead that certain specific violations of causal reasoning can helpfully be formulated, and integrated in a theoretically unified approach to the *post hoc*. To be sure, our analysis is only a start and depends directly on the assumption that causal language is coherent and open to logical analysis. The efforts of Suppes, Burks, Lewis, Domoter, the Victoria Group, and others to develop various elements of a theory of causality within modal, algebraic, and inductive frameworks give us some reassurance, and we are moved to suggest that much of the reluctance to speak of causes stems from an assumption that the causal relation is totally deterministic, that is, from (what we think is a fallacy) a conflation of causation with something like logical entailment [201]. A logically adequate essentially probabilistic explication of causal language should be welcomed by social scientists as a useful theoretical adjunct to experimental design and data processing techniques.

A more pressing need is the advancement of the study of informal fallacies beyond the inchoations and banalities of "The Standard Treatment" offered in introductory logic texts [225], [234], [222]. This is an area, crucially important to the pedagogy of logic and philosophy, that has been sadly understudied. Many of the stock examples have, remarkably, come down to us virtually intact from Aristotle. The traditional passing-on of unanalysed examples bereft of theory, accompanied by *ad hoc* systems of classification, has resulted in a body of lore that perpetrates rather more than it explicates the fallacies. Our analysis here is not the final word on the *post hoc*; we hope rather that it provides a systematic groundwork for the further exploration of causal fallacies.[18]

Chapter 10
Arresting Circles in Formal Dialogues

The study of the informal fallacies has an importance acknowledged by its place in both ancient and contemporary logic texts. Unfortunately, their study produces what Hamblin [80] has called the Standard Treatment, bereft of theory, laden with dogma and hoary illustrations appropriated from Aristotle, a corpus that often seems more to perpetrate the fallacies than to explicate them. Recently, however, a few writers have been seeking for some basic theoretical understanding of what these fallacies are, and we hope here to add something to this effort with some remarks about the *petitio principii*. Notwithstanding invitations occasioned by ambiguities which lurk in the title of this paper, our aim is not to report on dialogical circles we have known and loved, but rather to say something about how they may be stopped.

1. BACKGROUND

It is of some interest to determine the extent to which circular reasoning can be studied by means of three distinct kinds of system. They are: (i) systems of doxastic-epistemic logic such as may be found in the work of Kripke [117], Hintikka [98], and Wu [107]; (ii) systems of formal dialectic or formal dialogues as represented by Hamblin [80] and Lorenzen [126]; and (iii) such non-intensional, non-game-theoretic standard approaches as are afforded by first-order logic. De Morgan [50, p. 254ff.], presents the first systematic case for an approach to the *petitio* in the style of (iii), and a note of Hoffman[1] pursues a similar method. *De Morgan's Thesis*, as we shall call it, is that every genuine instance of the *petitio* is reducible to the case in which a conclusion is identical to some premiss or conjunctive part of a premiss. The trouble with this, however, is that examples abound of clearly circular arguments in the form of the disjunctive syllogism, yet in which the conclusion is not identical to any premiss or conjunctive part thereof.

It is ventured in Sanford [168] and in Woods and Walton [235] that De Morgan's treatment is inadequate to the *petitio*, and in Woods and Walton [234] that a satisfactory understanding of this fallacy involves, essentially, epistemic-doxastic notions. One example of an epistemic circularity condition, let us call it (CDE), might provide that an argument is circular in relation to a given person, if, and only if, in order to know that some premiss (or conjunctive part of a premiss) is true, the person to whom the argument is directed must know that the conclusion is true.

Still, at least one philosopher has felt (Barker [13]) that begging the question always presupposes a context of disputation, a setting in which there is a controversy over one or more issues, and that outside such a context, the *petitio* does not occur. Such a preference for a dialogical treatment of the fallacy probably arises from the suspicion that in Approach (i) there lie unacceptable presumptions of psychologism. After all, it might be protested (though without too much force we would suggest), the overtly 'subjective' parameters of Approach (i) are not likely to capture the essence of what is, let us remember, a logical matter.

However, it seems to us likely that the best account of begging the question will incorporate elements of all three approaches. In [234] and [235] we have pursued the *petitio* mainly from the perspective of (i), namely simple frameworks of doxastic logic, though there was also some informal attention paid to dialogical considerations. Here, however, our focus will be, in the manner of Approach (ii), more single-mindedly dialogical. Our task will be to determine whether the *petitio* can adequately be characterized in certain kinds of dialectical games: one constructed by Hamblin for this purpose; another, constructed by us following Kripke's semantics for intuitionistic logic; and still another, which is a simplification of the second.

2. THE GAME H

'Why-Because-System-with-Questions', designed by Hamblin [80, p. 265ff.], affords a possible solution of problems of organization of commitment. There are two participants, White and Black, each of whom has a commitment-store containing a finite set of statements, and each participant may add or delete commitments to or from his own store according to rules set out below. Let C be a's commitment-store, where a is a participant. Then, for one rather natural interpretation of the notion, one we have employed elsewhere, one might say that $x \in C$ iff $(\exists y)(Bay \wedge y \dashv x)$, with '$B$' the belief-operator. Let us say at once that this is *not* the interpretation that we wish to employ for Hamblin's game. For that purpose commitment-stores are finite, whereas on the current interpretation they are infinite whenever they are non-empty. The language of this game is basically first-order, restricted to a finite set of atomic statements. Axioms of the system are contained in the initial commitment-stores of both participants. White moves first, Black responds, and each continues in turn to make one move at a time. The capital letters $S,T,U,...$ are variables of the metalanguage for statements.

Locutions may consist of the following types:

(i) ⌜Statement S⌝ or, in certain special cases, ⌜Statements S,T⌝.

(ii) ⌜No commitment $S,T,..., X$⌝, for any finite number of statements $S,T,...,X$ (one or more).

(iii) ⌜Question $S,T,...,X?$⌝, for any number of statements (one or more).

(iv) ⌜Why $S?$⌝, for any statement S other than a substitution-instance of an axiom.

(v) ⌜Resolve S⌝.

Two categories of rules are given; locution rules and rules of commitment-store operation.

Locution Rules:

S1. Each speaker contributes one locution at a time, except that a 'No commitment' locution my accompany a 'Why' locution.

S2. ⌜Question $S, T, ..., X?$⌝ must be followed by
    (a) ⌜Statement-$(S \vee T ...\vee X)$⌝ ('-' for negation)
or (b) ⌜No commitment $S \vee T \vee ...\vee X$⌝
or (c) ⌜Statement S⌝ or
      ⌜Statement T⌝ or
      ——————— or
      ⌜Statement X⌝
or (d) ⌜No commitment $S, T, ..., X$⌝

146   Chapter 10

S3. ⌜Why S?⌝ must be followed by
    (a) ⌜Statement -S⌝
  or (b) ⌜No commitment S⌝
  or (c) ⌜Statement T⌝ where T is equivalent to S by primitive definition.
  or (d) ⌜Statements T, T ⊃ S⌝ for any T.

S4. ⌜Statements S, T⌝ may not be used except as in 3(d).

S5. ⌜Resolve S⌝ must be followed by
    (a) ⌜No commitment S⌝
  or (b) ⌜No commitment -S⌝ .

Commitment-store operation

C1. ⌜Statement S⌝ places S in the speaker's commitment store unless it is already there, and in the hearer's commitment-store unless his next locution states ⌜-S⌝ or indicates 'No commitment' to S (with or without other statements); or, if the hearer's next locution is ⌜Why S⌝?, placement of S in the hearer's store is suspended until the hearer explicitly or tacitly accepts the proferred reasons (see below).

C2. ⌜Statements S, T⌝ places both S and T in the speaker's and hearer's commitment-stores under the same conditions as in C1.

C3. ⌜No commitment S, T, ..., X⌝ deletes from the speaker's commitment-store any of S, T, ...,X that are in it and are not axioms.

C4. ⌜Question S, T, ..., X?⌝ places the statement S ∨ T ∨ ...∨ X in the speaker's store unless it is already there, and in the hearer's store unless he replies with ⌜Statement -(S ∨ T ∨ ....∨ X)⌝ or ⌜No commitment S ∨ T ∨... ∨ X⌝ .

C5. ⌜Why S⌝ places S in the hearer's store unless it is there already, or he replies ⌜Statement -S⌝ or ⌜No commitment S⌝.

It is of some importance that in (H) commitment-stores are not closed under the classical logical operations. Hamblin considers requiring that the statements in a commitment-store be *consistent* [80, p. 263f.] but rejects this because "consistency presupposes the ability to detect even very remote consequences of what is stored, and this would itself make nonsense of certain kinds of possible dialectical application." On the other hand, Hamblin suggests [80, p. 264] that certain "very immediate" consequences of a commitment might also be regarded as commitments in a given system. The general idea therefore seems to be that (H) can be regarded as a base system upon which closure requirements of various strengths might be built up for various purposes of application.

There is another important general aspect that calls for comment in understanding the motivation of (H); the retraction of commitments by a participant is allowed. That is, statements may be deleted from, as well as added to commitment-stores at appropriate moves. Particularly difficult questions of degrees of closure of commitments under logical operations concern retraction. What is to happen if a participant retracts commitment to T or even replaces it by $-T$ when he is committed to both S and $\ulcorner S \supset T \urcorner$? Again, Hamblin would have it that specific rules need to be laid down in specific systems to deal with this sort of situation.

Hamblin in defending the base system (H) stresses that a commitment is not to be thought of as a 'belief' of the participant who has it, and he disavows any implication that the interest or point of commitment-stores is psychological. It is well to notice also that [80, p. 260ff.] develops a formal version of the *Obligation Game* which has no provision for retraction of commitments. It is clear from this treatment that interesting formal games of dialogue *without retraction* can be constructed. We should also point out that while such games might have wide and various applications, Hamblin's primary purpose is to reveal structures of argument that might throw some light on forms of argument relevant to the study of the traditional informal fallacies.

3. CIRCLE GAMES

Hamblin discusses in [80, p. 268ff.], various modifications and additions to both sets of rules for various purposes, but since our interest here is in the representation of circular reasoning, two rules are of special importance. The first is a rule for when the 'Why?' proposer is regarded as inviting his opponent to convince him:

(W) $\ulcorner$Why $S?\urcorner$ may not be used unless S is a commitment of the hearer and not of the speaker.

Otherwise 'Why?' is academic. The second rule is specifically designed to block circular reasoning:

(R1) The answer to ⌜Why S?⌝, if it is not ⌜Statement - S⌝ or ⌜No commitment S⌝, must be in terms of statements that are already commitments of both speaker and hearer.

Following Van Dun [186, p. 110], dialogues will be represented by means of diagrams (see also Stegmüller [179] and Lorenzen [126]). The column on the left indicates the moves of White, those on the right the responses of Black. Pairs of moves are numbered on the left, and sequences of moves are set out using the method of nested sub-diagrams of tableaux.

The two most elementary forms of circular reasoning realizable in H can be represented by these dialogues. A, B, and C are atomic statements,

WHITE	BLACK
(1) Why A?	Statements $A, A \supset A$.

WHITE	BLACK
(1) Why A?	Statements B, $B =_{df} A$
(2) Why B?	Statements A, $A =_{df} B$.

Some other sequences, not mentioned by Hamblin, also represent kinds of circular argument.

WHITE	BLACK
(1) Why A?	Statements B, $B \supset A$
(2) Why $B \supset A$?	Statements A, $A \supset (B \supset A)$.

WHITE	BLACK
(1) Why A?	Statements B, $B \supset A$
(2) Why B?	Statements A, $A \supset B$.

The latter sequence represents a paradigm of circular argument, and in the sequel we will call an argument having this form *a circle game*. In a circle game it is possible to have a third step, C, intervening between the beginning and the end of the circle, as follows.

Arresting Circles in Formal Dialogues 149

WHITE	BLACK
(1) Why A?	Statements B, $B \supset A$
(2) Why B?	Statements C, $C \supset B$
(3) Why C?	Statements A, $A \supset C$.

Of course, the same form of sequence can be carried through to n steps, allowing for many intervening steps before the circle is closed by Black.

WHITE	BLACK
(1) Why A?	Statements A_1, $A_1 \supset A$
(2) Why A_1?	Statements A_2, $A_2 \supset A_1$
.	.
.	.
.	.
(k) Why A_{n-1}?	Statements A_n, $A_n \supset A_{n-1}$
(k+1) Why A_n?	Statements A, $A \supset A_n$.

The second and third sequences we looked at above may be likewise expanded. Though analogous things could be said about these other two kinds of games, in what follows we confine our remarks to circle games.

4. ADEQUACY OF H + (W) + (R1) FOR REPRESENTING THE PETITIO

How do (W) and (R1) block the *petitio*? Consider a two-statement circle game (the fourth game we looked at above). When Black responds, ⌜B, $B \supset A$⌝ at step (1), it is required by (R1) that both statements be in the commitment-store of both Black and White. Thus by (W), White's move at (2) is illicit, since he may ask ⌜Why B?⌝ only if B is not a commitment of his.

However, it would appear that an interesting form of sequence can be constructed without violence to H + (W) + (R1) but which has instances that may plausibly be interpreted as circular. Let us see. For the reader's convenience we set out the initial commitment-store of each participant in brackets at the head of the tableau. A superscript indicates at which step an addition is made; a stroke indicates deletions; and a superscript at the head of the stroke marks the step at which that statement was removed from the store.

WHITE [$A \supset B, B \supset A, A^2, \not{B}^3$]   BLACK [$A, B, A \supset B, B \supset A, C$]

1. Why A?	Statements B, $B \supset A$
2. Statement A	Statement C
3. No commitment B; why B?	Statements A, $A \supset B$
4. Statement B	

Description: Black accepts the truth of A and B and also their equivalence (mutual material conditional). White accepts the mutual implication, and accepts B. White asks ⌜Why A?⌝ and Black responds by citing B and ⌜$B \supset A$⌝, each of which is accepted by both parties. White concedes A. Black then moves on to something else. But then White 'gets the jitters' and (see Rule (W)) retracts his commitment to B, asking ⌜Why B?⌝. Black responds by citing A and ⌜$A \supset B$⌝, both of which are now in the commitment-stores of both parties. White concedes B. Note that at step 3 White is inconsistent, i.e., his commitment-store contains A and $A \supset B$ while at 3 he retracts commitment to B. To be consistent, he should also retract commitment to A or $A \supset B$. However, he moves back to consistency at 4. This he may do: commitment-sets are not closed under logical operations.

One way of understanding the dialogical sequence of Section 4 is that White can know that A is true at 2 only on the basis of B, as set out in Black's response at 1. Then by (CDE) (see Section 1) White's knowing that B on the basis of A closes the circle. We now develop this idea using the intuitionistic semantics of Kripke [117].

The model Kripke uses is that of the *tree*, such as the example given just below,[2] where G is the origin, the unique bottom node, and H_1, H_2, H_3, H_4 are ascending nodes of the tree. A sentence letter, A, B, C, is verified at a point if written above that node; if omitted, it is unverified at that point. Thus in the example below, A is verified at G but B is unverified at G:

Arresting Circles in Formal Dialogues 151

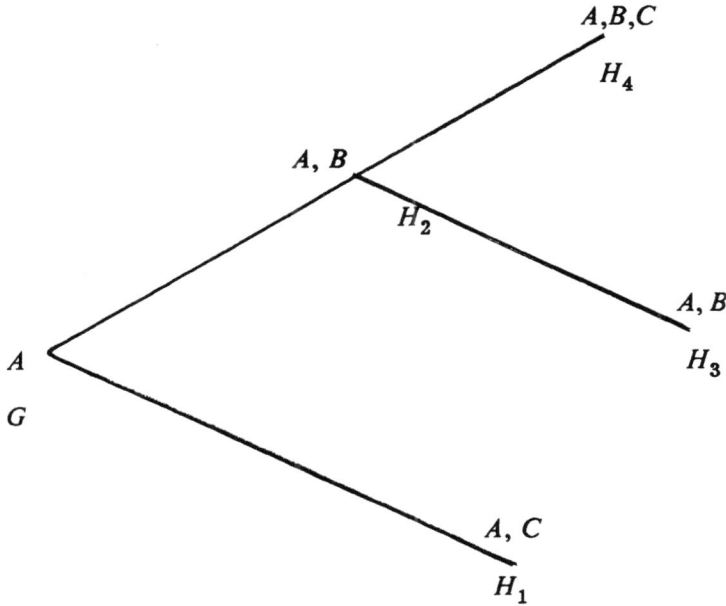

The nodes H_i represent points of time or 'evidential situations' at which times we may have various pieces of information. If we have enough information to ground a particular proposition A at a point H, we say A is verified at H; if we lack enough information to ground a proposition A at H, we say A is not verified at H.

The Kripke tree structure as a semantic representation of the idea of advancing states of knowledge may be illustrated with reference to the example above.[3] One possibility is that we may remain 'stuck' at G without gaining any new information. But it is possible that we will gain enough new information to 'jump' to point H_1 where we have a verification of C, as well as of A, or likewise to H_2, where we will have verified both A and B. Similarly, we could be 'stuck' at H_2 an arbitrarily long time, or advance to H_3 or H_4. Note that H_2 is significantly different from H_3 even though both A and B are verified at both points. As long as we remain at H_2, the possibility of advancing to H_4 remains open, but if we are at H_3 we have gained enough information to exclude ever verifying C.

What sorts of connectives do we have in a Kripke model? Disjunction and conjunction are exactly analogous to the classical truth-functions. Negation and implication however are quite different.[4] To assert $\ulcorner \neg A \urcorner$ at a point H, we need to know at H

not only that A has not been verified at H, but that A cannot possibly be verified at any later time, no matter how much new information is gained. Accordingly, ⌜¬A⌝ is said to be verified at H if, and only if, A is not verified at every point H' accessible to H. To assert ⌜$A \supset B$⌝ at H, we need to know that at any later situation, where we have a verification of A, we also have a verification of B. Thus ⌜$A \supset B$⌝ is said to be verified at H if, and only if, for every point H' accessible to H, A is not verified at H or B is verified at H'.

Formally, the model theory is as follows. An (intuitionistic) *model structure* is an ordered triple <G, K, R>, where K is a set, G is an element of K and R is a reflexive and transitive relation on R. An (intuitionistic) *model* on a model structure is a binary function $\emptyset (P, H)$, where P ranges over arbitrary proposition letters, and H ranges over elements of K, whose range is the set {T, F}, and which satisfies the condition: if $\emptyset (P, H)$ = T and HRH' (H, H' K), then $\emptyset(P, H')$ = T. We get from atomic propositions to formulae $A, B, C, \ldots$ of the propositional calculus by the following clauses.

(a) $\emptyset(A \wedge B, H)$ = T iff $\emptyset(A, H) = \emptyset(B, H) = T$; otherwise $\emptyset(A \wedge B, H)$ = F.

(b) $\emptyset(A \vee B, H)$ = T iff $\emptyset (A, H) = T$ or $\emptyset(B, H) = T$; otherwise $\emptyset(A \vee B)$ = F.

(c) $\emptyset(A \supset B, H)$ = T iff for all $H' \epsilon K$ such that HRH', $\emptyset(A, H')$ = F or $\emptyset (B, H') = T$; otherwise $\emptyset (A \supset B, H')$ = F.

(d) $\emptyset(\neg A, H)$ = T iff for all $H' \epsilon K$ such that HRH', $\emptyset(A, H')$ = F; otherwise $\emptyset (\neg A, H)$ = F.

A formula A of propositional calculus is called *valid* iff $\emptyset (A, G)$ = T for every model on a model structure <G, K, R>.

The notion that there is an epistemic priority in proving propositions may be represented in a Kripke model, and consequently, one kind of circularity can be expressed in the Kripke framework. Given an argument of the form '⌜$A, A \supset B$⌝ is verified at H'', where A represents the premiss-set and B the conclusion, we might say that the argument does not beg the question if, and only if ⌜$B \supset A$⌝ is not established at H. We oversimplify in that it is usually one premiss or part of a premiss, and not the whole premiss-set, that begs the question, but this complication presents no real difficulty. Thus our condition provides, in effect, that the conclusion of a non-circular argument must be established at some point in the evidentiary process after premissory verification. The condition may not be adequate to the fullest understanding of *petitio* in all respects - and we shall not attempt to show generally that it is (see [234]) - but it does give us a sufficiently compelling condition to make our demonstration of the incompleteness of Hamblin-games in an appropriate Kripke model of some interest to the concerns of this paper.

Consider now the following representation in a Kripke model of the dialogue of Section 4. By Black's move at 1, B is true at G, and B and A are true at H.

B	B, A
G	H

White's move at 2 is presumably to accept the truth of A and H. Up to this point argument $\ulcorner A, A \supset B \urcorner$ is non-circular, since A is established at H, a point later than G. But Black clearly violates the circularity condition at 3, since in this case,

B, A	B, A
G	H

whether the conclusion is A or B, in neither case is the conclusion established at a point earlier than the premiss. Thus in the Kripke model Black commits the fallacy of *petitio* at 3. Of course we assume throughout that White and Black intend to assert their various statements A, B, $\ulcorner A \supset B \urcorner$ and $\ulcorner B \supset A \urcorner$ at the same point, G. Otherwise, the fallacy is more of equivocation than *petitio*. It begins to appear, then that the game H + (W) + (R1) does indeed permit a fallacy of *petitio* on our version of the Kripke interpretation. We might also construe this result as additional confirmation of a conjecture of [234] and [235] namely, that *petitio* has the epistemic aspect of Approach (i), although, as we shall have occasion to say below, this remains very much a conjecture.

5. CUMULATIVENESS

Whether Hamblin's rules are in fact adequate to forestall all circular reasoning requires a closer examination of similarities and differences between the system (H) and Kripke's intuitionistic semantics [('K') let us say for short].

One outstanding difference between (H) and (K) is that the latter is essentially cumulative or incremental in a way that the former is not. On a Kripke tree, if a proposition is verified at a point (node) then that proposition must remain verified at every succeeding point (at every related node). Whereas in the game (H), if a statement is in the commitment-store of a participant at given a point (or move of that participant) it does not follow that the statement must remain in his commitment-store at every succeeding move he might make (at every next line of the tableau). (H) is not essentially cumulative because, as we have seen, (H) is the sort of game that, unlike the Obligation game or other games of dialogue, allows for retraction of commitments. Thus is would seem that any structure that consists of an ordered set of points can be thought of as having the property of essential cumulativeness or not, regardless of whether the points are interpreted as possible worlds or moves in a game, regardless of whether the framework is epistemic or game-theoretic. To define an appropriately general notion of cumulativeness, then, we need the following ingredients: (1) a set of points $w_i \in W$, (2) an ordering relation $<$ on the w_i, (3) a language L, a set of propositions or statements $A, B, C,...$, and (4) a function f that takes a pair $\{w_i, A\}$ onto a set $\{1, 0\}$. The idea behind (4) is that a given proposition can either obtain or not (have the value 1 or 0) at any given point w_i. Now the definition can be given: a system $<W, <, L, f>$ is *cumulative* if, and only if, for any two points $w_i, w_j \in W$, for any proposition A, if A has a given value (1 or 0) at w_i then A has the same value at w_j if $w_i < w_j$.

It is now easily seen that the Kripke system is cumulative whereas the Hamblin system (H) is not. The proofs in both cases are straightforward once the basic idea is sketched out, so we simply offer the sketches. First we show that the Kripke systems is cumulative. A relation $<$ defines what Kripke calls a *tree* over a set W if for any $w_i, w_j, w_k \in W$, if $w_i < w_k$ and $w_j < w_k$ then $w_i = w_j$. Then we let f correspond to Kripke's $\emptyset$ and L to Kripke's intuitionistic calculus defined by the conditions (a)-(d) for $\wedge, \vee, \supset$, and $\neg$ given in [117, p. 94]. Then we note that Kripke sets down for arbitrary proposition letters P and for his binary function $\emptyset$ $(P, H$, where H is the set of 'nodes' of the tree and $\emptyset$ corresponds to our f, the following condition: if $\emptyset (P, H) = T$ and HRH' $(H, H' \in K)$, then $\emptyset (P, H) = T$. Then he shows that once this property has been stipulated for a propositional letter (atomic formula) it follows for the complex formulae formed by clauses (a)-(d). So obviously Kripke's system is cumulative in our sense, for once a formula A is verified at a node H_k it will be verified (i.e., have the value T, corresponding to our value 1) at every node 'beyond' H_k in a given tree structure.

We show (H) to be non-cumulative by letting each w_i correspond to a line in the tableau of a dialogue. Then we can think of the lines ordered by the relation < on W, i.e., 'w_j is a later line than w_i'. Then we take a fixed participant, a, and we let $f(A, w_i)$ = 1 be satisfied in relation to a if the statement A is a commitment of a line a at line w_i of the dialogue. And we let $f(A, w_i) = 0$ be satisfied in relation to A if the statement A is not a commitment of a at line w_i of the dialogue. Obviously then there is no problem constructing a model of a dialogue permitted by the rules of (H) where A is a commitment of some participant at a given step, and then A is no longer a commitment of the same participant at some later step. Indeed, our problematic circle game of Section 4 provides one such example.

Now that we are equipped with a reasonably clear notion of cumulativeness, let us look back to the troublesome dialogue of Section 4. We can now pose the following formidable objection to the thesis that there is a circle in this dialogue according to the criterion of circularity constructed in the Kripke model in Section 5. The objection is this: there seems to be a circular sequence in the dialogue only if one sees it through the rose-coloured glasses of the essentially cumulative scaffolding of circularity typified by the framework of Section 5. The suggestion is that once we see that White retracts his commitment to B at 3, and that this is a legitimate move, the intuition that there is a circle in the dialogue quickly pales into unconfidence. That is, we may strictly speaking have a circle, but it would be nonsense to call it a fallacy. After all, in a non-incremental dialogical game, White has the right to retract his commitment to B at 3, and then change his mind *again* at 4. White could be accused of vacillating, but this does not amount to *petitio*.

This line of thought throws new light on the problem because the operative distinction changes from epistemic/dialogical to cumulative/non-cumulative. The critical factor in whether or not a *petitio* is committed is not whether the system is designed to model dialogical exchanges or epistemic states but whether it is cumulative in regard to certain values that the propositions (statements) are said to have.

6. GROUNDEDNESS

Another reason for thinking the dialogue of Section 4 is somehow circular is that one might naturally read White's statement that A at 2 is a *response* to the move of Black at 1 in the sense that White accepts *A on the basis that B*. Then of course later at 4, when White appears to accept *B on the basis that A*, the circle is closed. The fact is, however, that nothing in the rules of (H) tells us that a move like White's statement at 2 constitutes a response *on the basis* of a statement set out in a previous line. Hamblin, let us remember, disavows such psychological assumptions. Nevertheless, it is open to us to add a notion of groundedness to the dialogical steps of (H) in an effort to capture this intuition. Could we not define groundedness, for example, as follows: a statement *A is grounded on* a statement *B* in a Hamblin game if, and only if, *A* is a move of some participant at step *k* of a dialogue and ⌜*B, B* ⊃ *A*⌝ is a move of another participant at step *k*-1? Then a dialogue could be said to be circular if two statements are grounded on each other. By this criterion, our specimen in Section 4 is indeed circular. The relation of groundedness would presumably be transitive, and the relation of non-circular groundedness would be irreflexive and anti-symmetrical. Groundedness could also be defined on Kripke models as follows: *Q* is *grounded on P* at *H* in a Kripke model if, and only if, *P* is verified at some node *H* and *Q* is verified at any node *H″* such that (i) *HRH″*, and (ii) there is no node *H′* such that *HRH′* and *H′RH″* for distinct nodes *H*, *H′*, *H″*. In other words *Q* is grounded on *P* if *Q* is verified at the next node to one *P* is verified at. By the use of definitions of this sort, we seem to be able to reinforce the argument that at least one form of circularity in Section 4 can be identified.

Our own view is that while we are prepared to concede that something might be done with the notion of groundedness in the Kripke type of cumulative framework, we are hard-pressed to see what such a notion could amount to in a game that allows for retractions. The problem is, how do we deal with step 3 (taking our sequence of Section 4 again as example)? When White retracts *B*, are we still to say that his acceptance of *A* is 'grounded on' *B*? If his acceptance of *A* is grounded on something he now rejects, shouldn't he also reject *A*? Well, perhaps if he 'remembers' and if he hasn't discovered other grounds for *A* in the meantime, or if he had other reasons for accepting *A* all along that now disincline him to reject *A*. Each of these possibilities yields a new way of looking at groundedness, and each in turn can be expected to modify our intuitions as to whether the dialogue is circular. What we seem to have here is a host of new parameters concerning factors such as 'total evidence' versus 'partial evidence', factors that could be very elusive to define. In short, defining groundedness of *A* on *B* relative simply to the statement of ⌜*B, B* ⊃ *A*⌝ by the other participant at the previous step of a dialogue seems to us too thin a definition properly to support any single, clear intuition that the dialogue is circular or not.

7. A LINEAR SIMPLIFICATION

It is perhaps worth noting that in systems such as (K) one can have both strong and weak negation. This suggests that there are two distinct classes of possible connectives there definable: (a) the classical connectives, namely, ∨ and ∧ as defined by Kripke *plus* weak negation, i.e., ¬p at H has the value T if, and only if, p does not have the value T at H, and the classical conditional, $p \supset q$ at $H =_{df} \neg(p \wedge \neg q)$ at H; (b) the intuitionistic connectives, namely the conditional → and the negation ¬ defined by Kripke: (1) $p \rightarrow q$ is true at H if, and only if, q has the value T at every point accessible to p at H, (2) ¬p is true at H if, and only if, p has the value T at no point accessible to H. Notice that all the classical connectives are definable exclusively in reference to a single point H, whereas the intuitionistic connectives are defined in reference to a *spread* of points H_i. This is rather neat; for as long as we are at one single point (evidential situation) and are abstracting from the cumulative epistemic aspect, we have classical logic. But as soon as we bring in the epistemic aspect, involving the transition from a given point to a set of related points in the evidential progression, we get into a richer logic (which calls to mind, by the way, the analogy between IC and S4). In this richer logic we are essentially concerned with sets of points ≥2. Thus it doesn't so much matter what the individual connectives are as whether they can be defined exclusively in reference to a *single* point in the model. It is this latter property that seems to characterize their essential intuitionistness (or epistemicness, if you like). This brings out the importance of the notion of cumulativeness again, because 'cumulativeness' requires for its expression at least a pair of points. And it underlines why, in classical alethic logic, we do not get a very good model of arguing in a circle, or of the direction of evidentiary inference.

We can now easily enough see that the *petitio* can be characterized in an even simpler framework than that given by Kripke. For consider the model of a set of points ordered on a line:

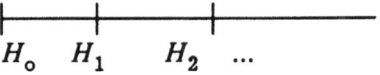

This structure is a Kripke tree with following condition: if H_iRH_j and H_iRH_k then $H_j = H_k$. In other words, it is a 'tree' but it is all trunk; no 'branching' is allowed (this is what mathematicians call a 'chain'). So suppose we have a set of points H_i ordered by a

Kripke R-relation on a line or chain as above. Consider the following sort of business:

```
  p      p,q     p,q,r ...
  ├───────┼───────┼──────────
  H₀      H₁      H₂ ...
```

Now we adopt the rule, in picking out arguments over such an array, that only the 'new' letter is picked as conclusion at each stage. At stage H_1, by this rule, p is the premiss-set and q the conclusion. Plainly there is no circularity here. Next, at H_2, $\{p,q\}$ is the premiss-set and r the conclusion. Again, no circularity here. And so forth. The structure is essentially cumulative in a way that systematically prevents circularity. At each step the 'conclusion' it new, relative to the previous 'evidentiary situation'.

Notice too that Kripke's conditions rule out 'backward reasoning', e.g.,

```
  p,q     p
  ├───────┼──────
  H₁      H₂
```

but not 'circular reasoning' (no real evidentiary progression), is ruled out.

```
  p       p
  ├───────┼──────
  H₁      H₂
```

However it is easy to ban circles by simply adopting the rule above. Such a rule corresponds with our suggestion above for banning circles in the Kripke model. But it is interesting to note that the cumulativeness (and consequently circularity) can be modelled linearly.

8. CONCLUDING REMARKS

It is not so much whether the system is epistemic or dialogical that tends to account for how well the fallacy of *petitio* can be modelled; the critical parameters would seem to pertain to such notions as cumulativeness and groundedness that might be defined in *either* type of system. Of course we have offered in these pages nothing like an exhaustive account of these notions, and it *might* turn out that the various categories have some well-defined correlations. For example, it might turn out that epistemic logic is best thought of as essentially cumulative and dialogical logic not. In the meantime, however, until more is known about the study and classification of these not very well defined families of systems, we shall have to declare the question open. What our work does suggest, however, is that instead of trying the two traditionally disparate methods of studying *petitio*, the epistemic and dialectical, more might be accomplished by studying their similarities and differences in a comparative way. We also think the model dialogue of Section 4 is an especially important specimen against which to test our theories. If nothing else, it is rich in intuitive conflicts.

It seems to us that, in modelling the *petitio*, cumulative systems are simpler. Consequently cumulative systems such as (K) yield a clear and relatively simple model of circular argument. When one extends this model to non-cumulative systems many complexities ensue, some of which may be expected to (and do) obscure the idea of *petitio*. Those who cleave to the relatively clear notion of circularity of Section 5 and who cannot bring themselves to understand the *petitio* in non-cumulative contexts may well persist in arguing that in the non-cumulative system (H) circles may occur. For it does seem clear that there is no effective way of prohibiting dialogues that, from a cumulative perspective on the *petitio*, are circular. Even so, the available evidence suggests that there are special difficulties in defining a workable concept of *petitio* adequate to non-cumulative systems. Whether such evidence should be regarded as chastening we do not venture to guess.

Still, there is some reassurance in recognizing that in the Kripke framework a useful theory of *petitio* can be formulated. We would suggest in as much as epistemic logic has an affinity for cumulative systems, we can have a theory of *petitio* that is epistemic in character, even if more exotic possibilities remain as potential competitors.

Chapter 11
The Fallacy of Ad Ignorantiam

This paper outlines a three-part analysis of the traditional informal fallacy of *ad ignorantiam*. As initially characterized, the fallacy consists in arguing that failure to prove falsity (truth) implies the truth (falsity) of a proposition.

First, the fallacy is located within confirmation theory as a confusion between the categories of "lack of confirming evidence" and "presence of disconfirming evidence". Second, the structure of the fallacy can be seen as an illicit negation shift in Hintikka-style epistemic logic. Third, the fallacy can be studied as an attempt to unfairly shift the burden of proof in a dialectical game. We suggest that research on *ad ignorantiam* needs a broadening of the scope of philosophical logic to encompass concepts of correct argument in these three contexts.

John Locke wrote in 1690 that men commit the fallacy of *ad ignorantiam* when requiring an adversary in debate to admit what they allege as a proof, or assign a better [125, p. 278]. If, however, the proof is a good one, it is not easy to see how there is anything fallacious in what Locke described. Irving Copi wrote in 1972 that men commit the fallacy of *ad ignorantiam* when they argue that a proposition is true simply on the basis that it has not been proved false [40, p. 76]. Obviously, however, sometimes failure to prove falsity *does* constitute some legitimate evidence for the truth of a proposition. The ease with which we can overturn both accounts of the fallacy suggests that ignorance or arguing from ignorance is a relic of our heritage that has been preserved intact. Still more disheartening, even these two clearest and most helpful articulations of *ad ignorantiam* appear to lack any real agreement or unity of direction.

Incoherence and lack of direction is the style we are accustomed to in the long and dismal history of the study of informal fallacies. Modern logic texts continue the tradition, sequestering the study of fallacies from the growing, better defined branches of logic. While the crucial pedagogical importance of the fallacies is acknowledged by their continued appearance on university curricula, teachers do not feel that they can commend the study of fallacies as a decisive strategy in the adjudication of argument. The bereftness of theoretical resources, poorly compensated for by the proliferation of hackneyed examples, does not encourage a systematic treatment of the basic indeterminancies which students quickly perceive in "The Standard Treatment' [80, ch. 1]. We would here like to address ourselves to a long-overdue initial attempt to locate a framework for the further analysis of the *ad ignorantiam*. We do not think we have the complete analysis of this fallacy, for, as we hope to show, this task partly awaits upon development in other, largely unresolved, areas of related study.

1. FIND THE FALLACY

It is part of the lore of experimental science that the want of confirming evidence for a hypothesis is not to be confused with disconfirming evidence; and stern warnings against this confusion are found in the usual homilies to scientific method. Conversely, so we are told,[1] the want of disconfirming evidence does not confirm a hypothesis either. Arguments having the form of one or the other of the following pair are said, therefore, to commit the fallacy of *ad ignorantiam*.

(A_1) There is no disconfirming evidence for H.
Therefore, H is confirmed.
(A_2) There is no confirming evidence for H.
Therefore, H is disconfirmed.

No doubt there is truth in these admonitions. The fact is however that in some contexts, arguments having the forms (A_1) and (A_2) are *not fallacious*. Consider the case where there are high antecedent probabilities in favour of H; then (A_1) may not be fallacious. Similarly, low antecedent probabilities might vindicate (A_2). Or if it seems to beg the question to bring in antecedent probabilities, consider the case where $\Pr(H) = .5$ (exactly). Even here (A_1) and (A_2) need not be fallacious. We might imagine a case where a series of ingenious experiments has been run, where any negative result would have occasioned the downfall of H, yet in which no negative results were in fact reported. *Ceteris paribus*, we might well suppose H to have been, to a certain degree, confirmed; and (A_1) would be regarded as inductively correct. Similarly, if deliberate, well-formulated attempts to confirm H were repeatedly to give rise to *nil* results, all else being equal, our

The Fallacy of 'Ad Ignorantiam' 163

confidence in H would quite reasonably be diminished, provided the antecedent epistemic probabilities are regarded as fixed. Either way, whether *a priori* probabilities are counted or not, (A_1) and (A_2) do not seem to deserve their reputation as fallacies. Can it be that *ad ignorantiam* is not a fallacy? Or is it rather that the schemata (A_1) and (A_2) do not manage to specify it?

2. CONFIRMATION: DIS AND UN

The fallacy might more perspicuously be located if we reminded ourselves that any testable hypothesis has a class of statements (deductively or inductively) deducible from it (plus some auxiliary hypotheses) that we might call the *test class*.[2] For a hypothesis to be of any legitimate experimental interest, the test class must exclude certain statements (such as those that are neutral for the purposes of confirmation or disconfirmation), and it must include certain other statements, although we may not know whether the statements it includes are true, or are likely to be true, false, or likely to be false. Thus for any hypothesis we require the general totality of statements to be divisible into two disjoint, non-empty classes, the test class, and the neutral class.

For any hypothesis, H, any statement may be said to be either H-neutral or H-testable, but not both. If we were to imagine the class of all statements as a line, a hypothesis H would cut off a segment of that line representing the H-testable statements.

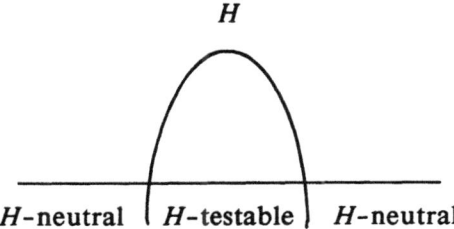

Let us also agree that hypotheses can be confirmed or disconfirmed in varying degrees, and that some hypotheses are neither confirmed nor disconfirmed: call these *unconfirmed*. Accordingly there are three disjoint classes of hypotheses.

| unconfirmed | disconfirmed | confirmed |

That there is some such tripartite partitioning essential to confirmation theory is a relatively uncontroversial assumption. How the distinction is to be implemented in actual experimental practice is of course a very different and substantive problem of confirmation theory, and we will not try to resolve it here.

Three further minimal assumption are needed that seem essential to any reasonably standard notion of confirmation. First, we assume that if an H-testable statement is confirmed, then H is either confirmed or disconfirmed. Second, we assume that if an H-testable statement is disconfirmed, then H is either confirmed or disconfirmed. That is, we will not require that H-testable confirmation always leads to H-conformation or H-testable disconfirmation to H-disconfirmation. Nor will we indicate the manner in which the degree of confirmation of H-testable statements statistically escalates or lowers the degree of confirmation of hypotheses. But third, we do assume that if an H-testable statement is unconfirmed, then H is also unconfirmed, prior probabilities aside. Thus our three assumptions are these: unconfirmation is carried over from H-testable statements to H, and the union of disconfirmation and confirmation is likewise carried over.

3. A FALLACY OF CONFIRMATION

We see from the assumptions of the previous section that the following equivalences obtain.

(E_1) H is confirmed $\leftrightarrow$ $\neg$(H is unconfirmed v H is disconfirmed)
(E_2) H is disconfirmed $\leftrightarrow$ $\neg$(H is confirmed v H is unconfirmed)
(E_3) H is unconfirmed $\leftrightarrow$ $\neg$(H is confirmed v H is disconfirmed)

We have it, therefore, that the pair of argument forms below are valid.

(V_1) H is not disconfirmed
        Therefore, H is confirmed v H is unconfirmed
(V_2) H is not confirmed
        Therefore, H is disconfirmed v H is unconfirmed

But the following pair are incorrect.

The Fallacy of 'Ad Ignorantiam'   165

(F₁)   *H* is not disconfirmed
       Therefore, *H* is confirmed

(F₂)   *H* is not confirmed
       Therefore, *H* is disconfirmed

The latter pair are not valid forms of argument, since the conclusion fails to exhaust all (in this case both) possible alternatives. Comparison of (V₁) with (F₁) and (V₂) with (F₂) shows that the fallacy in both cases consists in suppression of the possibility that *H* may be unconfirmed, i.e., the live possibility that there are no *known* data for *H* is omitted. And in so saying, we have the suggestion *ignorantiam* has an epistemic aspect.

4. AN EPISTEMIC FALLACY

Typically the "proved" in 'Nobody has ever proved that ∅ is false (true), therefore ∅ is true (false)' carries an epistemic load; the *ad ignorantiam* therefore can be significantly viewed as an epistemic fallacy. The argument here may exhibit one of the following pairs of epistemic structures.³

(F₃) $\neg (\exists x) Kxp$ (F₄) $\neg (\exists x) Kx \neg p$
       Therefore, $\neg p$ Therefore, p

Simply because nobody knows that p is true, it does not follow that p is false. Simply because nobody knows that p is false, it does not follow that p is true. Thus (F₃) and (F₄) are fallacies. Moreover, it is easy to see that (F₃) and (F₄) are invalid argument forms in Hintikka's system, or any standard system, of epistemic logic. For in such systems, for any a ⌜$Kap \supset p$⌝ and ⌜$Ka \neg p \supset \neg p$⌝ are theorems, but ⌜$\neg Ka \neg p \supset p$⌝ and ⌜$\neg Ka \neg p \supset p$⌝ are not theorems. The fallacy can thus be exhibited as confusion between the pair,

(F5) $\neg Kap$ (V5) $Ka \neg p$
       Therefore, $\neg p$ Therefore, $\neg p$

or between the pair.

(F6) $\neg Ka \neg p$ (V6) $Ka \neg \neg p$
       Therefore, p   Therefore, p

The forms on the left are incorrect and the ones on the right correct, in standard systems. The resemblance within each pair is easy to see when the scope of the negation operator is made explicit as above. The fallacy is effected by an unlicensed shift of the negation sign to the left of the K-operator, a manoeuvre not difficult to obscure in ordinary speech.

5. A DIALECTICAL FALLACY

An important class of fallacies involve constructions having to do with the onus of proof.[4] In general, if a person asserts a sentence in the context of argument, he may reasonably be obliged to support it with some independent evidence. This is plain enough, yet often in the course of a complex or convoluted debate, considerations of onus of proof may become blurred. The simplest scenario is where we might have a debate between Mr. X and Mr. Y, and Mr. X at some point in the debate asserts that p. Then later in the debate, perhaps when the issue has been somewhat clouded, Mr. X may aggressively demand that Mr. Y produce evidence of the negation of p, in a case where Mr. Y has expressed or implied disbelief that p obtains. Now it is altogether reasonable for Y to admit that he has no definite or articulatable positive evidence to prompt us to the negation of p, but yet on the other hand, that it is not obvious to him that the evidence sufficiently warrants the acceptance of p either. In other words, Mr. Y may maintain that no relevant evidence sufficient to favour either acceptance or rejection is available. In this case it may be quite unreasonable, even fallacious, for X to insist that Y produce evidence for p's negation. Since Mr. X originally asserted that p, it would seem that the obligation is his to marshall evidence in p's behalf. Alternatively, it would seem appropriate that X retreat into a more neutral modality concerning p.

None of these requirements can be warranted by a standard, monolectical logic, however powerful; standard logics do not recognize, much less proclaim upon, the dialectical patterns of propositional assertion and denial. Only a dialectical logic, of the kind studied by Hamblin, more reminiscent of games theory than standard logic, [80, ch. 8], [233] can chart the reasonableness of those skeptical challenges and refutations. A *dialectical system* is a regulated dialogue or family of dialogues with two or more participants who speak in turn in accord with a set of rules or conventions. Formal dialectic consists in the setting up of simple systems of logically well-formulated but not necessarily very realistic rules, and studying dialogues that can be run in accordance with the rules. Hamblin has developed several interesting formal dialectical systems.

A dialectical game that begins to capture the *ad ignorantiam* suggested by the simple scenario above requires a rule allowing the questioner to put forward ⌜Why p?⌝ without requiring that his question be open to the same sorts of justification procedures covering the case of the respondent who opts to assert one of the pair {p, ¬p} in response to a question. For this simplest case, the kind of mechanism required is illustrated by a syntactical rule in Hamblin's 'Why-Because-Systems-with-Questions' (p. 265f.):

(S3)   'Why S?' must be followed by
         (a) 'Statement ¬S'
or   (b) 'No commitment S'
or   (c) 'Statement T' where T is equivalent to S
             by primitive definition
or   (d) 'Statement T, T ⊃ S' for any T.

Transparent attempts to shift the burden of proof unfairly, as in the scenario we considered earlier, can now be classed as violations of S3. For example, the following specimen of dialogue between *a* and *b* is barred by syntax rule S3.

b : Why S?
a : Why ¬S?

It should also be added that not all questions of burden of proof are so easily dealt with. We need to distinguish kinds and degrees of assertion, commitment and concession, if the formal system is to reflect actual argumentation more adequately to the more refined violations of burden of proof. But the simplest cases of *ad ignorantiam* can be usefully understood as syntactical aberrations of dialectic.

6. THE ANALYSIS

We suggest that an adequate understanding of the fallacy of *ad ignorantiam* is to be achieved only through recognition of its three principal features. First, it is a fallacy of confirmation; it is an argument from the unconfirmed to the disconfirmed. Second, the argument from ignorance has an epistemic aspect: it is the fallacy of arguing from the not-known to the known-to-be-false. Third, the *modus operandi* of the fallacy in actual argumentation can only be understood as a dialectical deviancy. The *ad ignoratiam* is committed when at least one of these three breaches is committed. Sometimes *ad ignorantiam* will specifically consist in exactly one of these errors. But more often, two or more of them will be present, and in varying degrees.

Each of the above three aspects of the *ad ignorantiam* outruns standard first-order logic. Accordingly, further research on *ad ignorantiam* requires an enlarged notion of correct argument for such contexts.

7. PRESUMPTIONS

It is often observed that there is one special context where *ad ignorantiam* is not a fallacious mode of reasoning, namely in the courts. Does not the law rule, for example, that a person is presumed innocent until proven guilty? Thus, writes Copi, "[t]he defense can legitimately claim that if the prosecution has not proved guilt, this warrants a verdict of not guilty" [40, p. 77]. There is no fallacy here, however, according to our analysis of *ad ignorantiam*, for it is no fallacy to *presume* a statement is false, unless presumption is meant to imply knowledge of the falsity or disconfirmation of the statement.

The legal requirement is not that innocence be *confirmed* or *known*, but only presumed. Moreover, proof of innocence in law is not an epistemic achievement. Proof of innocence is merely a presumption undefeated. Such presumptions are not dialectically fallacious either, for they can be incorporated into dialectical games as legitimate responses through S3 by the utilization of the device of *common commitment-stores* for a set of participants. In general, the set of commitment-stores for a number of participants may be disjoint - that is, have no statements in common - but there may be some statements that occur universally, in the commitment-store of all participants; these may be regarded as common or universal presumptions of all parties. In general, as Hamblin shows, any dialectical game requires certain metalogical agreements, concerning procedures and rules, but object-statement presumptions or common commitments are also possible. Where there exists a commitment, C, universal to the stores of all participants, C, in effect becomes a legitimate substitution instance for T in option (d) of S3. C operates here like a presumption in actual argumentation. It can operate within syntactically sanctioned procedures for responding to ⌜Why S?⌝ , thus evading the charge of *ad ignorantiam*, without making a knowledge claim or confirmation-assertion concerning the substance of the argument.

Presumptions have a genuine logical role to play, not only in the law, where they are extremely prominent, but also in scientific pursuits. It has often been observed that well-established hypotheses in science have a certain inertia, or presumption of correctness, that a revolutionary hypothesis must strive to overcome by a strong preponderance of evidential backing. The more well-entrenched the current hypothesis, the more strongly must its would-be competitors be supported by evidence. Of course the unjustified use of such presumptions is a source of fallacies - the *ad verecundiam* and *petitio principii* may occur here as well as the *ad ignorantiam*. But legitimate presumptions are possible in dialectical contexts, and they need not always involve the *ad ignorantiam*. The structure of dialectical games can permit them, and in fact they reflect their legitimacy.

8. TESTABILITY

A complication often introduced into the discussion of *ad ignorantiam* in many texts, however incidental to the structure of the fallacy as we see it, is the question of verifiability. Typical examples of *ad ignorantiam* have to do with ghosts, telepathy or other psychic phenomena, or religious argumentation, all contexts where questions regarding the verifiability, the testability-in-principle of hypotheses, naturally arise. *Ad ignorantiam* here takes the form: nobody has been able to devise an experimentally adequate means of testing H, therefore H is false, or disconfirmed. Rather, however, the suggestion is often implicit in such arguments that the hypothesis in question has never been confirmed because it is not a testable hypothesis at all - in our terms, the class of H-testable statements is either null or maximal, excluding no statement, and consequently what we have is merely a pseudo-hypothesis. Perhaps the fallacy suggested here is confusion of a testable hypothesis with a pseudo-hypothesis of one kind or another - but this is not the fallacy of *ad ignorantiam* as we conceive it.

However, a genuine case of *ad ignorantiam* can occur in these contexts (if fallacies are ever genuine) as follows: nobody has ever been able to formulate a testable hypothesis that concerns the existence of ghosts, therefore ghosts do not exist. Even granting the premiss, this argument proceeds from the unknown to the known-to-be-false, from the unconfirmed (though not through lack of data, as such, but through lack of theory) to the disconfirmed. It is thus a true case of the classical *ad ignorantiam*. We have here a variant of the fallacy, but one that conforms to our analysis.

9. ADDENDUM

In some recent work [235], [226], [232], we have found that the Kripke semantics for intuitionistic logic can be very useful in providing models for forms of argument appropriate to the study of informal fallacies that have an epistemic character. Above we have emphasized that *ad ignorantiam* should be thought of as an epistemic fallacy, and we have suggested that the fallacy can be modelled in a Hintikka-style syntax of epistemic logic. [98] Therefore it seems reasonable to think that *ad ignorantiam* might be studied by the Kripke semantics [117] too. Indeed this possibility is stimulated even further by the well-known mappings between the modal system S4, which is isomorphic with Hintikka's epistemic logic, and the intuitionistic calculus. [136] We proceed to outline the gist of the Kripke semantics below and briefly indicate how the notion of *ad ignorantiam* reasoning can be modelled in it.

The model Kripke uses is that of the *tree*, such as the example given just below, [117, p. 97ff.] where G is the origin, the unique bottom node, and H_1, H_2, H_3, H_4 are ascending nodes of the tree. A sentence letter, A, B, C, is verified at a point if written above that node; if omitted, it is unverified at that point. Thus in the example below, A is verified at G but B is unverified at G:

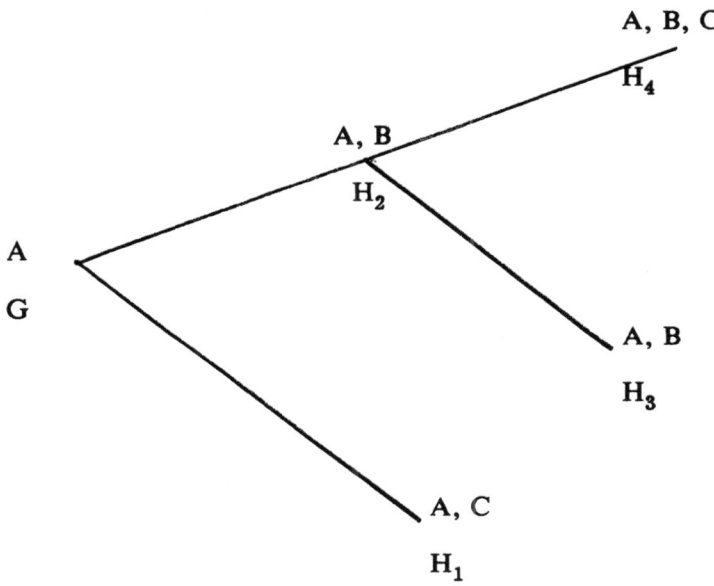

The nodes H_i represent points of time or "evidential situations" at which times we may have various pieces of information. If we have enough information to ground a particular proposition A at a point H, we say A is *verified* at H; if we lack enough information to ground a proposition A at H, we say A is *not verified* at H.

The Kripke tree structure as a semantic representation of the idea of advancing states of knowledge may be illustrated with reference to the example above [117, p. 99]. One possibility is that we may remain "stuck" at G without gaining any new information. But it is possible that we will gain enough new information to "jump" to point H_1 where we have a verification of C, as well as of A, or likewise to H_2, where we will have verified both A and B. Similarly, we could be "stuck" at H_2 an arbitrarily long time, or advance to H_3 or H_4. Note that H_2 is significantly different from H_3 even though both A and B are verified at both points. As long as we remain at H_2, the possibility of advancing to H_4 remains open, but if we are at H_3 we have gained enough information to exclude ever verifying C.

What sorts of connectives do we have in a Kripke model? Disjunction and conjunction are exactly analogous to the classical truth-functions. Negation and implication however are quite different[7]. To assert $\ulcorner \neg A \urcorner$ at a point H, we need to know at H not only that A has not been verified at H, but that A cannot possibly be verified at any later time, no matter how much new information is gained. Accordingly, $\ulcorner \neg A \urcorner$ is said to be verified at H if, and only if, A is not verified at every point H' accessible to H. To assert $\ulcorner A \supset B \urcorner$ at H, we need to know that at any later situation, where we have a verification of A, we also have a verification of B. Thus $\ulcorner A \supset B \urcorner$ is said to be verified at H if, and only if, for every point H' accessible to H, A is not verified at H or B is verified at H'.

Formally, the model theory is as follows. An (intuitionistic) *model structure* is an ordered triple <G, K, R >, where K is a set, G is an element of K and R is a reflexive and transitive relation on R. An (intuitionistic) *model* on a model structure is a binary function ∅ (P, H), where P ranges over arbitrary proposition letters, and H ranges over elements of K, whose range is the set {T, F}, and which satisfies the condition: if ∅ (P,H) = T and HRH' (H, H'ε K), then ∅ (P, H') = T. We get from atomic propositions to formulae A, B, C, ... of the propositional calculus by the following clauses.

(a) ∅ (A ∧ B, H) = T iff ∅ (A, H) = ∅ (B, H) = T; otherwise
    ∅ (A ∧ B, H) = F.
(b) ∅ (A ∨ B, H) = T iff ∅ (A, H) = T or ∅ (B, H) = T; otherwise
    ∅ (A ∨ B) = F.
(c) ∅ (A ⊃ B, H) = T iff for all H'ε K such that HRH',
    ∅ (A, H') = F or ∅ (B, H') = T; otherwise ∅ (A ⊃ B, H') = F.
(d) ∅ (¬A, H) = T iff for all H'ε K such that HRH', ∅ (A, H') = F; otherwise ∅ (¬A, H) = F.

A formula A of propositional calculus is called *valid* iff ∅(A, G) = T for every model on a model structure <G, K, R>.

We should notice that it is possible to distinguish between strong and weak negation in the Kripke model. For *strong negation*, ¬A, it is required that A be not-verified at *every* point accessible to a given point, whereas for *weak negation*, ~A, A is thought of as simply having the opposite value of A exactly at the given point. ~ is thus in effect classical negation, and can be defined exclusively with reference to a single point. ~ is intuitionistic negation, and characteristically requires essential reference to a *spread* of points in its definition. For ¬A to obtain at H, ¬A must obtain at every H_i accessible to H. Thus, the inference form '~A, therefore ¬A' is in general fallacious, and indeed clearly represents a form of *ad ignorantiam*; merely because A is unverified at a given point in time it need not follow that A cannot be verified at any future time. In terms more reminiscent of our Hintikka-style analysis: merely because A is not known to be true it does not follow that A is known to be false, i.e., can never be known to be true.

Obviously much more remains to be said about the various interrelationships between the Hintikka and Kripke logics in this connection. However, we must leave the elaboration of our conjecture to another paper. For the moment we merely wish to present the basic idea of how intuitionistic logic can provide us with a method for approaching the epistemic aspect of the basic structure of *ad ignorantiam* by means of an exact formal procedure. *Ad ignorantiam* begins to appear somewhat less obscure than informal fallacies are commonly thought to be.

Chapter 12
Circular Demonstration and Von Wright-Geach Entailment

Students of the literature on entailment [215], [18] are familiar with the von Wright-Geach definition of entailment [190], [73]: p entails q if, and only if, by means of logic,[1] it is possible to come to know the truth of $\ulcorner p \supset q \urcorner$ without coming to know the falsehood of p or the truth of q.

A competitor of the Lewis account of entailment as strict implication, the von Wright-Geach definition has enjoyed little success. Perhaps one reason is that the definition is essentially epistemic, and entailment is not widely thought to be an epistemic notion.[2] The notion of entailment aside, however, the von Wright-Geach notion does intriguingly seem to be applicable to one interesting epistemic concept of obvious logical interest. In this article, we suggest that the von Wright-Geach *definiens*, in effect, partially defines the concept of a *non-circular demonstration*, and we argue that the von Wright-Geach *definiens* can be extended in a natural way to yield a full definition for non-circular demonstration. Thus we think of the latter notion in a frankly epistemic way. We have argued in [235] and [234] that circularity of argument (*petitio principii*) is best thought of as an epistemic matter, and for those who agree with us on this point it may not seem too surprising that there is a connection between circularity and von Wright-Geach "entailment". For those who disagree with our thesis that *petitio* is essentially epistemic, we hope that establishing the connection in question may serve to diminish the disagreement. In either case, we think that this new application of the von Wright-Geach framework is interesting in its own right.

1. CIRCULARITY CONDITIONS

In [234] we argued that the history of the subject, which has largely followed in the tradition set by Aristotle in various remarks in the *Topics, De Sophisticis Elenchis*, and the *Rhetoric*,[3] suggests the wisdom of recognizing two broad types of circularity-conditions, neither of which however is as theoretically well-

176 Chapter 12

behaved as we might like. According to the *equivalence conception* an argument is circular when the conclusion is equivalent to (or even, in some version, identical to) some premiss.[4] According to the *dependency conception*, an argument is circular where some premiss depends on the conclusions, i.e., where one cannot know that the premiss is true without knowing that the conclusion is true. Both types of conditions have been stated in both an epistemic and also a purely alethic idiom, but we cite the epistemic variant of the dependency-type condition to fit the framework of von Wright [190]. An epistemic dependency condition (C) in the style of von Wright would read as follows: an argument, 'p therefore q' is non-circular if, and only if, it is possible to come to know the truth of p without coming to know the truth of q. One may know the truth of q, as a matter of fact, but one must have some *other* means of knowing the truth of p - that is, some means independent of q.

2. THE VON WRIGHT-GEACH DEFINITION

How does (C) fit the von Wright-Geach definition of entailment? We can find the answer by recognizing that (C) needs to be modified because there is another kind of circularity that can occur where classical connectives are used.[5] The material conditional ⌜$p \supset q$⌝ has the property of being true where the consequent, q, is true. But if one were to propose an argument of the form *modus ponens* on the basis that the conditional premiss ⌜$p \supset q$⌝ obtains because the conclusion q obtains, one would have committed a blatant dependency-*petitio*. We could diagram this state of affairs as follows:

$$\begin{bmatrix} \{p \supset q \\ \phantom{\{}p \end{bmatrix} \longleftarrow \\ \longrightarrow q $$

Not only are we obliged to pass from the premisses to the conclusion as intended, but we are likewise obliged to pass from the conclusion to a premiss, thus closing the "circle". Now the von Wright-Geach definition requires that it must be possible to come to know the truth of ⌜$p \supset q$⌝ without coming to know the truth of q, thus preventing this form of circularity from arising. Thus (C) and the von Wright definition seem to complement each other. Can they be put together to define something like perhaps 'non-circular entailment'?

3. REVISING THE DEFINITION

To accomplish this, two major differences of orientation need to be smoothed out. First, the von Wright-Geach definition, unlike (C), is restricted to what we come to know "by means of logic". Second, (C) is concerned with how we come to know the truth of p (the premisses), whereas the von Wright-Geach definition merely requires the possibility that we come to know the truth of ⌜$p \supset q$⌝ without coming to know the falsehood of p. Plainly if we are to have a concept of non-circular argument that avoids both kinds of dependency-circularity mentioned above, we will have to take into account the question of whether it is possible to know the truth of p without coming to know the truth of q. So we will smooth out the second difference by broadening the von Wright-Geach definition to take into account the possibility of coming to know the truth of p. On the first difference however, we will narrow the notion of argument to arguments that are concerned with "the means of logic" alone, i.e., we will consider only demonstration (proof), that species of argument where p and ⌜$p \supset q$⌝ are possible to come to be known true by means of logic,[6] where the premiss-set and conclusion consist exclusively of theorems.[7] Thus we propose a definition of non-circular demonstration. We do not, however, wish to reject the possibility that the sort of definition we offer may be extended to arguments where premisses cannot come to be known to be true by means of logic.

Now it is interesting to observe that if we put (C) and the von Wright-Geach definition together as they stand, an inconsistency is yielded. The conjunctive definition, suitably modified as suggested in the previous paragraph, would read: for theorems φ and ψ, ⌜$\varphi \supset \psi$⌝ is a *non-circular demonstration for* ψ if and only if, by means of logic, it is possible to come to know the truth of φ and ⌜$\varphi \supset \psi$⌝ without coming to know the falsehood of φ or the truth of ψ. But, given what von Wright adopts as axioms (as shown below), if it is possible to know that p is true then p is true. Likewise, if it is possible to know that p is false, then p is false. Thus the phrase "the falsehood of φ or" is not only redundant in the definition, but is actually the occasion of its inconsistency. Accordingly, the fully modified definition may be given as follows: φ, ⌜$\varphi \supset \psi$⌝ is a *non-circular demonstration for* ψ if, and only if, by means of logic, it is possible to come to know the truth of φ and ⌜$\varphi \supset \psi$⌝ without coming to know the truth of ψ. We think this "definition" gives a quite satisfactory account of dependency-circularity in the context of demonstration, and that its closeness to and interdependence with the von Wright-Geach definition of entailment is what gives the latter its plausibility. Yet whether our "definition" really provides a helpful or adequate analysis of circularity of demonstration depends ultimately on what sort of account we can give of the three undefined terms that occur in it: "demonstration", "it is possible to come to know the truth of φ", and "without". Von Wright has some interesting suggestions to make in this regard, and we will comment on these briefly.

4. REFINEMENTS

According to von Wright, the following two equivalences are true: 1. *p* is *demonstrable (provable)* if, and only if, it is possible to come to know the truth of *p* by means of logic, and 2. *p* is *demonstrated* if, and only if, it is possible to demonstrate *p*. Thus he thinks it important to distinguish between a proposition's being demonstrable and its being demonstrated. The latter entails the former, but the converse does not obtain.

The term "possible" in the second equivalence is meant by von Wright in the rather unusual way that if it is possible to come to know that *p* is true, then *p is* true. In a perhaps more usual sense of "possibility", if *p* is false we would say that it is possible that it might have become known to be true simply because (if it is a contingent proposition) it is possible that it might have been true. But in von Wright's (unusual) sense of "possible", if *p* is false it is *impossible* that it should become known to be true. This new sense of "possible" is the key to the analysis we propose of a non-circular demonstration. We would conjecture, although we are not yet very confident about it, that "possible" here is not used in the usual sense[8] but in the context of coming to know the truth of a *theorem* (a non-contingent proposition). On our account of the matter, von Wright [190, p. 186] proposes what amount to the following axioms for D*p* (*p* is demonstrated and MD*p* (*p* is demonstrable).

(A1) $D\varphi \supset \varphi$
(A2) $MD\varphi \supset \varphi$
(A3) $D(\varphi \wedge \psi) \equiv (D\varphi \wedge D\psi)$
(A4) $MD(\varphi \wedge \psi) \equiv (MD\varphi \wedge MD\psi)$
(A5) $D\varphi \supset MD\varphi$
(A6) $D\varphi \supset DMD\varphi$ (demonstrated $\supset$ demonstratedly demonstrable).
(A7) $D\varphi \supset MDMD\varphi$ (demonstrated $\supset$ demonstrably demonstrable).

Finally, concerning "without", the third undefined term, von Wright proposes the following equivalence: φ is demonstrable independently of ψ if, and only if, it is possible that φ is demonstrated and ψ is undemonstrated. Thus given **D** and **M** (for possibility) understood after the fashion of von Wright, we can express our definition of 'φ, ⌜$\varphi \supset \psi$⌝ *is a non-circular demonstration for* ψ (i.e., $\varphi, \varphi \supset \psi \dashv \psi$) as follows:

$$\varphi, \varphi \supset \psi \dashv \psi =_{df} M(D\varphi \wedge D(\varphi \supset \psi) \wedge \neg D\psi).$$

No essentially epistemic concept is purely alethic (i.e., truth-theoretic or proof-theoretic); therefore an epistemic analysis of entailment is misdirected. We think that entailment is, so to speak, a "purely formal" matter. But we do not think that the notion of a non-circular demonstration is a purely formal matter,[9] but rather better viewed epistemically or information-theoretically.

However, we do not wish to become enmeshed in fruitless debate over what is or is not "purely formal". What we hope to have shown is that non-circular demonstration is at least not a purely subjective matter, and is indeed open to analysis.

Chapter 13
Laws of Thought and Epistemic Proofs

A common reaction among idealist philosophers to the classical syntactic characterization of proof so crisply articulated by Tarski [183], [182] is an urgent but inchoate *Angst* that something momentous is missing, an awesome intimation of bereftness. The simple fact is that in many pursuits proof involves an empirical appeal, an operation that Tarski (unlike Bradley and other idealists) excludes from the domain of proof and assigns to the company of *confirmation*. In Tarski's terms, empirical statements never even admit of the predicate *true*, let alone *proved*, unless perhaps they hanker after truth or proof in the limit in some dimly understood and always unsuccessful way. Some sources of queasiness about the syntactic account are not hard to pinpoint. On the syntactic account, (a) the injunction "Prove it!" can never be fulfilled, or even well-formedly uttered, in regard to an empirical statement, (b) ⌜∅, therefore ∅⌝ is valid for any well-formed ∅ in a given first-order language, (c) for any theorem, t, ⌜∅, therefore t⌝ is an impeccable proof for any well-formed ∅ in the classical languages. It is easy to think, with Bradley, that there is an important and legitimate, if not formally well-articulated sense of *proof*, characteristic of much of philosophical reasoning and argumentation, which is badly misrepresented by the syntactic conception. The kind of proof often sought after in theology, and sought after and found in the law, and the experimental and social sciences, is not purely syntactic. Idealists, of course, have frequently expressed dissatisfaction with the Tarski conception of *truth* [162]. We will here extend this skepticism to the Tarski notion of *proof*.

Perhaps in the end it will prove wiser to respect Tarski's Platonic sentiment that proof ("conclusive" proof) is never found in the social sciences or the law, but rather that such hypotheses are confirmed and such arguments probabilistically weighed.

But this is not a position we are prepared to start with. We will investigate the question whether a frankly epistemic conception of proof such as that favored by Bradley and Bosanquet [23] is initially feasible. Thus Bradley [25] argues for a broad sense of "demonstration" that would transcend the narrower mathematical idea captured by the standard syntactical conceptions of "proof".

Demonstration in logic is not totally different from demonstration elsewhere; proof is only one kind of demonstration. Logicians however seem generally not to be aware of this fact. When the mathematician "demonstrates" a conclusion the logician feels uneasy, though he cannot deny that the conclusion is proved. But uneasiness becomes protest and open renunciation when he attends at the "demonstrations" of the anatomist. He shudders internally at the blasphemous assertion that "this which I hold in my hand" is "demonstrated." But his trials are not over; the illiterate lecturer on cookery overwhelms him by publicly announcing the "demonstration" of an omelette to the eyes of females. But I think the logician has no real cause of quarrel even with the cook. For demonstration is merely pointing out or showing; and if the conclusion of an inference is seen and thus may be shown, so also may a nerve or again an omelette.

We ourselves appropriate the broader notion in the hopes of finding a concept of proof or demonstration suitable for the underlying theory of the major informal fallacies. As we have argued elsewhere, [233] it happens that from a purely syntactic point of view, the epistemic character of many of the fallacies is critically obscured if not altogether ignored, and this fact alone plays a role in the current arcane state of informal logic, sequestered from the developed branches of logic, pedagogically and practically important yet still in its theoretical infancy [225]. We will touch on this theme later with special reference to the informal fallacy of *petitio principii* [234].

1. MOOREAN PROOF

Moore is famous for his "refutation" of idealism. But, ironically, his favorite concept of proof is no stranger to the mainstream of idealism through its broadly epistemic orientation. Consider a set of premisses, P, and a unary set, C, a conclusion. An argument, "P, therefore C", may be said to constitute a successful *Moorean proof* for some individual a just where (1) a knows that P is true, (2) a knows that P deductively implies C, and (3) the argument is non-circular. The third condition is added to forestall the objection that "P, therefore P" (and various other dubious forms of proof) would qualify as a successful Moorean proof. The addition of (3) reveals that it is a *desideratum* of Moorean proof that it be possible that there are other ways of coming to know the truth of a statement than by proof, e.g., by direct observation. A celebrated instance of a successful Moorean proof is given in "Proof of an External World" [139], where Moore proves (or thinks he proves) the statement "Two hands exist". This he does by raising his right hand, saying "Here is one hand", and raising his left saying, "Here is another".

P1: Here is one hand.
P2: Here is another hand.
C: Two hands exist.

The argument constituted a successful Moorean proof for the observer of Moore's actions who knew that both premisses were true by direct observation provided he knew that P1 and P2 deductively imply C (and also provided that the argument is not circular).

There are a number of ways of explicating condition (3) of which, we think, the most plausible and appropriate is the following. An argument is *noncircular* for a just where it is possible for a to know the premisses are true without his knowing that the conclusion is true. [234] Thus "P, therefore P", for example, is generally if not always a circular form of argument because it is not possible in such cases for a person of even a minimal degree of rationality to know that the conclusion is true without knowing that the premiss is true. Of course the general and quite benign assumption that consigns "P, therefore P" to the carousel is this:

$(\forall a)(aK)(p \equiv p)$;

that is, everyone knows that a proposition implies and is implied by itself. We now turn to other epistemic assumptions.

2. EPISTEMIC ASSUMPTIONS

Our next minimal epistemic assumption is the principle that if a knows that p is true then a believes that p is true.

(4) $aKp \supset aBp$

Then, too, there is the requirement that if a knows that p is true then it must follow that p is true.

(5) $aKp \supset p$

In the minimal conception here outlined, knowledge involves psychological and alethic factors; however, the converses of (4) and (5) do not obtain, since in the standard view which we seek to preserve, knowledge should not collapse into belief *simpliciter* nor into truth *simpliciter*. It is also worth remarking that (5) embodies a rationality assumption for, since all truths are consistent (in standard modal accounts of consistency), it follows that everything anyone knows is consistent with everything else he knows.

(6) $(\forall p)\ (\forall q)\ (\forall a)\ (\ (aKp\ \&\ aKq) \supset M(p\ \&\ q)\ )$

Recent work shows that additional rationality assumptions beyond (5) can have surprising consequences that tend to deviate in interesting ways from common conceptions of knowledge [107], [99], [97], [108]. For example, do I know all the logical consequences of the statement that I know?

(7) $(p \rightarrow q) \supset (aKp \supset aKq)$?

No, it would seem that (7) exacts too severe a price - to know anything whatever even in these domains in which one is a relative simpleton, one must be logically omniscient. However, the weaker condition (7) seems less treacherous,

(8) $aK(p \rightarrow q) \supset (aKp \supset aKq)$,

and we might tentatively accept it. Accordingly, we put it that if there is a successful Moorean Proof for a and p, then a knows that p is true. No doubt conditions both stronger and weaker than (4), (5), or (8) might reflect a variety of more or less interesting conceptions of knowledge. For current purposes, however, we will relativize Moorean proof to the view of knowledge minimally characterized by the assumptions, (4), (5), and (8). So characterized, proof is seen as partly epistemic, partly psychological, and partly alethic. Trivially, then, Moorean proof outranges classical proof-theoretic, syntactic, purely alethic models of proof.

3. NON-CLASSICAL CONNECTIVES

Classically, any tautology is deductively implied by any statement whatever. Accordingly, a successful Moorean proof for any tautology is embarrassingly easy to produce. Consider some fairly complex tautology, t. Now consider some individual a, who knows that t is implied by "Flin Flon is in Manitoba" and who knows that the latter statement is true. For this individual, "Flin Flon is in Manitoba, therefore t" is a successful Moorean proof.

According to a widely held view of deductive implication, any necessary statement is deductively implied by any statement whatever. Similarly, we have the consequence of this view that any necessary statement can be furnished with a successful Moorean proof by yielding up some unrelated truth as premiss. These problems raise the question of whether the relation denoted by " $\rightarrow$ " in (8) can correspond to strict implication, or whether a nonclassical account of implication, such as relevant implication, [3], [2], [212] might not be needed for Moorean proof. One way of preserving the classical approach in the face of this problem would be to restrict C in "P, therefore C" to contingent statements. However, this leaves one with the rather empty feeling of *ad hocery* and with the nagging question, "Why *should not* necessary truths be susceptible of Moorean proof?"

However, it is interesting to note that the adoption of a relevant logic may also exclude noncontingent premisses or conclusions. For assume that a set of statements is minimally inconsistent if and only if it is inconsistent but no proper subset of it is inconsistent. Then, after Lehrer, [121] we could define a relevant deductive argument as one in which the conjunction of the premisses with the denial of the conclusion is minimally inconsistent. It would then follow, as Lehrer has shown, that no relevant deductive argument has a premiss or conclusion that is either tautological or inconsistent. Thus, like it or not, the contingency of premisses and conclusion is guaranteed by a standard view of relevance logic.

We ourselves prefer to deal with these embarrassments for Moorean proof, not in non-standard accounts of implication, and not in *ad hoc* simple-minded favoritism toward contingency, but rather by settling for a workably non-standard account of the connectives such as may be found in intuitionistic logic. A useful and illuminating link between epistemic logic and intuitionistic logic is suggested by the demonstration of a close structural similarity between S4 (essentially the Hintikka system for epistemic logic) [98] and IC, the intuitionistic calculus [151] by McKinsey and Tarski in 1948 [136]. (See Hughes and Cresswell [105, p. 306] for the details of this demonstration and the axioms of IC in a brief outline.) [105] An account promisingly well-suited to our needs covering Moorean proof is given by Kripke in his epistemic semantical analysis of intuitionistic logic [117]. We recapitulate briefly the gist of this analysis.

The model Kripke uses is that of the *tree*, as in the example just below, [117, p. 97f] where G is the origin, the unique bottom mode, and H_1, H_2, H_3, H_4 are ascending nodes of the tree. A sentence letter, P, Q, R, is true at a point if written above that node; if omitted, it is false at that point. Thus in the example below, P is true at G but Q is false at G:

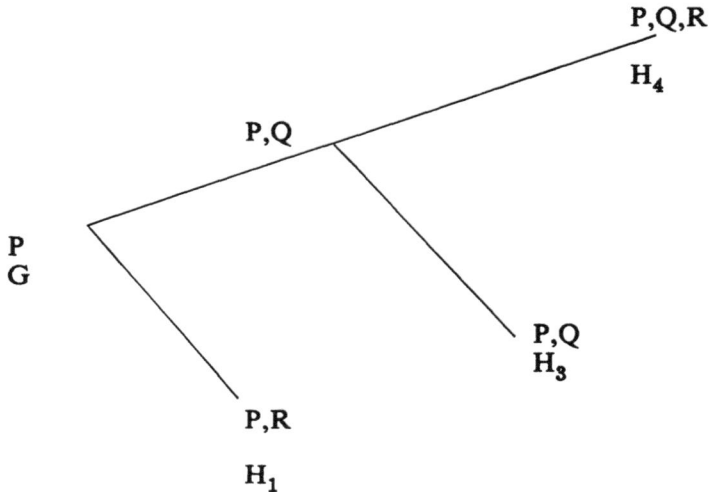

The nodes H represent temporally ordered 'evidential situations' at which times we may have various pieces of information. If we have enough information to prove a particular proposition P at a point H, we say P is *true* at H; if we lack enough information to prove a proposition P at H, we say P is *false* at H. "True" and "false" are not thought of in this context in their ordinary sense-it would be better to think of these terms as meaning, respectively, "verified" and "not verified". Thus when we say, "p is true at H" we mean in a Kripke model what might be expressed as "p has been verified to be true at the time H". Where we say, "p is false at H" in a Kripke model, however, we do not mean to assert that p has been *proved false* at H, but only that p has not yet been proved at H, but may be established later.

A Kripke tree-structure semantically represents the idea of advancing states of knowledge. This may be illustrated with reference to the example above [117, p. 99]. One possibility is that we may remain "stuck" at G without gaining any new information. But it is possible we will gain enough new information to "jump" to point H_1 where we have a proof of R, as well as of P, or likewise to H_2, where we will have proved both P and Q. Similarly, we could be "stuck" at H_2 for an arbitrarily long time, or advance to H_3 or H_4. Notice that H_2 is significantly different from H_3 even though both P and Q are verified at both points. As long as we remain at H_2, the possibility of advancing to H_4 remains open, but if we are at H_3 we have gained enough information to exclude every proving R.

What sorts of connectives do we have in a Kripke model? Disjunction and conjunction are exactly analogous to the classical truth-functions; but negation and implication are non-truth-functional [117, p. 99 & 94]. To assert '¬p' at a point H, we need to know at H not only that p has not been verified at H, but that p *cannot* be verified at any later time, no matter how much new information is acquired. Accordingly, '¬p' is said to be true at H if, and only if, p is false at every point H' accessible to H. To assert 'p ⊃ q' at H, we need to know that in any ensuing situation in which we have a proof of p, we also have a proof of q. Thus 'p ⊃ q' is said to be true at H if, and only if, for every point H' accessible to H, p is false at H' or q is true at H'.

4. INDIRECT PROOF

Moorean proof, as it stands, seems unable to accommodate the type of proof known as the *reductio ad absurdum*, for it is required by condition (1) for Moorean proof that the premisses be true. Yet it is far from uncommon for a proof of a conclusion, C, to proceed by demonstrating that the negation of a statement known to be true follows from the denial of C (*reductio ad falsum*). A special case of this indirect mode of proof, *reductio ad absurdum*, derives an inconsistency from the denial of C. The difficulty with absorbing these indirect modes of proof in the Moorean model is that indirect proofs not only do not require premisses that are known to be true, but they do not seem to require premisses at all. This difficulty can be circumvented, however, if we regard as a premiss the denial of the (known) true statement that is derived from C. Thus the defining conditions of a *successful Moorean indirect proof* would require, in addition to (1) and (3) as before, that (2'); *a* knows that ¬C deductively implies ¬P.

Whether *reductio ad absurdum* can likewise be incorporated in the Moorean model depends on the assumptions discussed above regarding the question of classical connectives. We can see this as follows. Classically, an inconsistent statement deductively implies any statement whatever. Now consider some individual, *a*, who knows that some inconsistency, ¬t, deductively implies "Flin Flon is not in Manitoba", a statement *a* knows to be false. For *a*, "Flin Flon is in Manitoba, therefore *t*" is a successful indirect Moorean proof. But should it be? Suppose *t* is Hartogs' Theorem. Would the above procedure be considered a successful proof for Hartogs' Theorem, in the sense of *proof* we seek to capture? If not, we are faced by the same alternatives as before if we wish to preserve the assimilation of the *reductio ad absurdum* to the Moorean model, namely (a) jettison the classical connectives for Moorean proofs, or (b) exclude necessary statements from the domain of Moorean proof.

5. CIRCULAR PROOFS

Elsewhere, we have argued that a version of condition (3) that is reasonably faithful to the traditional intuitions about the *petitio principii* can be achieved only by recognizing the essentially epistemic environment of this fallacy [234], [235]. An important epistemic dependency condition, (CDE), provides that an argument is circular for a person a where, in order to know that some premiss is true, a must know that the conclusion is true. (CDE) requires, in effect, that for an argument not to beg the question it must be possible for the person to whom the argument is directed to know that all the premisses are true without basing his knowledge on the conclusion, that is, without having to know that the conclusion is true. In short, the following must obtain, for each premiss P.

(9)   M (aKp & ¬aKC)

For example, in advancing an argument of the form,

(10) P_1: If B then C
        P_2: B
               Therefore, C

we normally expect that justification of P_1 and P_2 would be independent of C. If the justification of P_1 consisted in the claim that C entails P_1, the argument would be blatantly and unpersuasively circular. Non-circular use of *modus ponens* offers us the option of avoiding a conclusion-based knowledge-claim for any premiss. Thus *modus ponens* can be circular or not depending upon the epistemic context of its assertion. Other forms of argument, in virtually excluding knowledge of the premisses not based on the conclusion seem always, or nearly always, circular. For example,

(11) P_1: P
               Therefore, P

Here it is scarcely possible for anyone to know that the premiss is true without knowing that the conclusion is true. What seems required in our epistemic logic to assure the circularity of all arguments of the form (11) is the assumption of the axiom,

(12)   ¬(aKp & ¬aKp)

Even (12) will not always doom (11) to constant circularity, however, provided we allow that a successful proof for C for a may be achieved even where a already knows that C is true.

The argument may still be non-circular where a has the option of concluding that C on the basis that he knows that P implies C and he knows that P is true independently of his knowledge of C. It would seem to be desirable that our account of proof should allow for proving something we already know to be true. Alternatively, we might require that proof can proceed only from the more to the less well-known, but this seems an artificial requirement, and we shall not pursue it here.

Now, an acceptor of (11), even one of whom it is true that aK(P → P) applies, might still, it seems have some grounds for accepting the premiss of (11) other than the conclusion, P itself. But here is the rub. The independent grounds, in the form, say, of another statement, P', scarcely requires the intermediary of the premiss, P, to support the conclusion, P. For minimally, it must be the case that

(13)   aKp ≡ aKp

and hence it is epistemically indifferent whether we bring P' to bear on the conclusion of (11) or the premiss.

P_1: P because P'
       Therefore, P

is fallacious because we might as well eliminate the premiss, P, and bring down P' to bear directly on P as conclusion.

The issue emerges somewhat more clearly in the case of *modus ponens*. If we assume (8) plus

(14)   aK(p → (if q then p))

we have it as a consequence that, in regard to (10), if a knows that C is true, a knows that P_1 is true. a's knowledge here does not doom arguments of the form *modus ponens* to irredeemable circularity in all contexts for, if a is aware of an entailment from

C to P_1 is may be possible for *a* to avoid attributing his knowledge of P_1 to C, and in this case there need be not circularity.

6. CONCLUSIONS

Just as we suspect that many intuitionist criticisms of classical logics can be usefully seen as tacit appeals to epistemic conceptions of proof, so perhaps might the relevance of relevant logics and Kripke semantics for intuitionistic logics find a fruitful field of application in idealistic Moorean "proofs". We suggest that Moorean epistemic proof fills a large gap. Moreover, since we think we have shown that Moorean proof is at least initially amenable to analysis through assumptions in epistemic logic by means of non-standard logics, we propose Moorean proof as an interesting object of further investigations. We leave as a yet unsolved problem the fuller explication of (CDE) through further investigation of the modal expression in (9). Our hope is that M can be shown to be essentially epistemic in such a fashion that an adequate account of the *petitio principii* within an epistemic logic may be forthcoming. In the meantime, we commend the study of Moorean proof to all those of an idealist orientation who are friendly to notions of proof, argument, and inference of a kind that transcends the narrow syntactic conception. This includes, we suggest, virtually *every* academic discipline or methodology with the possible exception of certain branches of mathematics not concerned at all with contingencies. Thus the expression "laws of thought" once thought to be an idealist confusion of psychology and logic[1], an obstacle to progress of pure formal logic, may be shown to have a legitimate sphere of application as an object of philosophical and logical inquiry in the wider sense.

Chapter 14
What Type of Argument is an Ad Verecundiam?

Elsewhere [229], [241] we have stressed the need, in teaching informal logic, to include in the logical repertoire the skill of discerning the *type* of argument that the student is to evaluate. For if there is more than one type of argument, as we believe, the correctness or incorrectness of an argument may very with the factor of type. For example, if there are inductively correct arguments, some of them (perhaps even all of them) may be deductively incorrect (invalid). Consequently, neglecting this type of distinction could spawn many a fallacy. For example, a systematic sophist might take one's correct inductive arguments and rule them deductively incorrect, *ergo* bad arguments. For all their deductive incorrectness they may be perfectly good arguments, taken as what they were meant to be, i.e. inductive arguments. Thus the sophist's ploy is based on a true premiss of deductive incorrectness, but it is a sophistical refutation because it equivocates on the factor of type.

1. TYPE OF ARGUMENT

This factor of type is particularly critical in teaching the evaluation of arguments *ad verecundiam*. It has sometimes been thought reasonable that appeals to authority can be a legitimate type of argument - that is, not always fallacious - but rather fallacious only given that certain conditions of the appropriateness of the appeal fail to obtain. [225] Even so, one may ask - what the type of argument is involved? Hamblin [80, p. 218] suggests that we could start from the valid argument, "Everything X says is true, X said that p, therefore p", and expect to find weaker but still not fallacious forms of argument where premisses of the form "X is an authority on facts of type so-and-so" lend some support to p. Salmon [167, p. 64] asserts however that the appeal to authority is not deductively valid, for the premisses could be true and the conclusion false - no authority, by these lights, is infallible or omniscient. Rather, according to Salmon the appeal to authority may be inductively correct if it has this form; "The vast majority of statements made by X concerning subject S are true. p is a statement made by X concerning subject S; therefore p is true".

Who is right? Is the *ad verecundiam* a type of argument that can be either deductive or inductive, is it perhaps inductive but never deductive as Salmon urges, or could it be something else altogether, neither deductive nor inductive in character? These are fundamental questions for anyone who would want to find ways of teaching students to identify and evaluate the *ad verecundiam*.

2. TWO CHARACTERISTICS

Two fundamental characteristics of appeals to authority should be brought forward at this point. First, *ad verecundiam*, like its partner in crime *ad hominem*, is *subject-based*. That is, what one authority X asserts may in general be different from or even contradictory to what is asserted by another authority Y. Second, *ad verecundiam* is *subject-matter-sensitive*. That is, an authority's pronouncement that p may be correct or not depends on whether or not the subject-matter of p is one in which the putative authority is indeed a legitimate expert. Since neither the subject-based or subject-matter-sensitive characteristics are true of the standard or classical approaches to the logic of either deductive implication or inductive conditionals, it seems reasonable to think that there may be some deeper reasons why the *ad verecundiam* can be neither deductive nor inductive as a type of argument. But how could it be proved?

3. NEITHER INDUCTIVE NOR DEDUCTIVE

We would now like to introduce the thesis that arguments *ad verecundiam* could be of a type that is neither inductive nor deductive, and suggest that the required type is that of the *plausible inference* of Rescher [161]. Plausible reasoning comes to bear on cases of informational-overdetermination, e.g. inconsistency, where we have too much information and have to decide what must be given up. Characteristic therefore of the case of plausible reasoning is the less than total veracity of our sources, for in an inconsistent pair of pronouncements, one source must be wrong. In this climate, neither deductive nor inductive inference is *à propos*, and in fact Rescher proves that the required type of argument can be neither deductive nor inductive.

What Type of Argument is an Ad Verecundiam? 193

Here are the essentials of the proofs given in Rescher [161, p. 2ff.]. If the inference "X (a generally veracious but imperfect source) maintains p, therefore p" were deductively valid, then so would the following inference be deductively valid for some other generally veracious but imperfect source Y: "Y maintains ¬p, therefore ¬p". But if both inferences are indeed deductively valid then from "X maintains p" and "Y maintains ¬p" it follows that p ∧ ¬p is true. Clearly this consequence is absurd however, for merely because authorities maintain conflicting pronouncements it hardly follows that p ∧ ¬p is in fact true. The same consequence follows by elementary laws of probability from taking "X (a generally veracious but imperfect source) maintains p, therefore p is highly probable" as a correct inference [161, p.3]. Thus Rescher has shown that for an essentially subject-based (for two sources X and Y, or greater than two) appeal to authority, the type of inference can be neither deductive nor inductive. In essence, these disproofs reflect the conception that for multiple authorities that are imperfect and may be expected to have conflicting pronouncements, deductive and inductive models of inference are "too perfect". Hamblin's and Salmon's conceptions of the *ad verecundiam* are too idealized to adequately represent the practice of appeals to imperfect authorities whose pronouncements may clash. But confronted by contradiction we must not give up - even though deductive or inductive logics give no further guidance - but press on to resolve the contradiction by means of plausibility theory.

4. PLAUSIBLE ARGUMENT

Now that we have eliminated the deductive and inductive models, and identified plausible inference as a preferable model for the type of argument exemplified by the *ad verecundiam*, it would seem the way is open to an analysis of this fallacy. And so indeed it may be, but this is not a project we shall attempt here. Suffice it to say for the moment that as Rescher conceives it plausible inference is not subject-matter-sensitive, so at very least plausibility theory will have to be conjoined to a theory of the subject-matter content of propositions [200] in order to be adequate to the full *ad verecundiam*. These refinements aside however, we are at least in the position now of being able to identify one noteworthily insidious form of the *ad verecundiam*.

The fallacy we allude to occurs where an appeal to authority is construed so strongly, or such a lack of specification of its type of argument has transpired, that the argument is taken to have (a) deductive, or (b) inductive correctness. Yet if the appeal is meant to be taken - as it should be generally - to a less than perfectly veracious authority, then its construal as (a) or (b) is fallacious. The specific fallacy here lies not in the appeal to authority as such, but in the spurious escalation of the appeal towards a claim to a source of truth that is more perfect or infallible than a plausible argument has any logical right to be. In short, this fallacy is to misidentify the type of argument.

This particular error is of course not the only way in which an appeal to authority can go wrong, and elsewhere [229], [241] we have suggested that *ad verecundiam* is an umbrella concept for several specific pitfalls of argument from authorities. But this particular species of the *ad verecundiam* is an important one, we think, in teaching students how to confront and deal with the fallacies, because it underscores the need to take into consideration identification of the type of argument as a necessary skill of informal logic. The first step in attempting to adjudicate any allegation that a fallacy has been committed is to ask the question "What (exactly) is the argument?" Answering this question involves more than simply specifying a set of propositions - as in the approach of formal logic - it includes, among other tasks, specification of the type of argument that has been advanced.

Chapter 15
Equivocation and Practical Logic

Among those who deal with this sadly understudied area of logic, authors of introductory logic texts often seem to commit, more than explain, the informal fallacies. What the student gets is the Standard Treatment, [80,ch. 1] as Hamblin calls it, long on titillating examples and short on serious explanations or general guidelines. Not that the subject is unimportant, or even uninteresting. The study of fallacies is obviously of central concern and methodological relevance for philosophy and for the role philosophy plays in the humanities [221]. Students very often exhibit a lively initial interest in the fallacies, for it is easy to see their practical utility in evaluating everyday argument [230]. But the fact is that we lack theory [80], [233]. Discussion of the most rudimentary text-book examples quickly makes it plain that a non-arbitrary sorting of the 'correct' from the 'fallacious' argument is simply not available to us.

We will here attempt the beginnings of an approach to theory, and this will occupy us for the first five sections of the paper. We will, in particular, deal with the fallacy of equivocation, concentrating first on certain underlying issues of theory, in hopes that a foundation for the further fruitful study of this fallacy may be set down. We need, for one thing, to understand why and how equivocation can be seen, in theory and practice, as an *informal fallacy*. Accordingly, we shall begin by specifying certain conditions on what can constitute an informal fallacy.

We want to work towards the long-term objective of providing the logic student with working guidelines for the recognition and understanding of the fallacy of equivocation in actual argumentation; and this we pursue in the last half of the paper. This project has a strongly practical aspect, as we will see again and again. It means dealing with a collection of 'informal' or 'practical' difficulties that stand in the way of constructing a working theory of equivocation. Hence, obviously enough, it is an exercise in the metatheory of informal rather than formal logic. The essentially practical nature of the project may be conveyed by the following observations. An equivocation is an argument that turns on an ambiguity. Hence all we seem to need is a theory of meaning for the given language that we are working with, a theory that effectively processes ambiguities in the language, whereupon the main problem perhaps the whole problem of equivocation is solved. It would seem to follow that the primary problem is that of finding a general theory of meaning (or syntax, if our worry extends to the related fallacy of amphiboly). Would that this were so. However, contrary to optimistic expectations, we will find that equivocation in practice, that is to say, from the point of view of a working logic, is not as nicely behaved as these hopes suggest. Our investigations will indicate that a practically adequate doctrine of equivocation requires several kinds of 'pragmatic' extensions of a general theory of meaning. We have come to believe that the real problem deals not with which general theory of meaning is chosen, but how it can be extended to cope with these practical problems of argumentation. Therefore our strategy will be to concentrate on working out the practical adequacy conditions which any general theory of meaning would have to meet before beginning to qualify as a basis for a working theory of the fallacy of equivocation. Our problem is one of integrating theory and practice; we seek for practical guidelines. Thus we will not propound a general theory of meaning in this essay, or even disclose what our favourite theory is. We do hope however that the practical deliberations will encourage others to integrate semantical research into meaning with these pragmatic considerations.

1. THE NATURE OF THE PROBLEM

Generally there are thought to be two components of an informal fallacy: (i) it must be an incorrect argument, and (ii) it must be a kind of argument that might seem correct. In William of Sherwood's terms, the aspect of *non-existentia* (non-existence, incorrectness) is as much a problem in understanding equivocation as is the aspect of *apparentia* (semblance, seeming-correctness) [210, p. 134, note 10]. Some comments on both requirements are in order. Firstly, most of the traditional informal fallacies fall outside the explanatory capacity of first-order logic; so it is often difficult to specify the logical structure of the misdemeanour in question. As an example, such traditional fallacies as the *petitio principii* and the *ad ignorantiam* can be adequately understood only as deficiencies with respect to an essentially epistemic concept of argument, yet epistemic logic is certainly not first-order logic [234], [238], [235]. Nevertheless, even if the specification of the concept of argument involved might be non-classical, it remains reasonable to require that a fallacy be an *incorrect argument*. It would appear, then, that study of informal fallacies has a *bona fide* logical component and should not be expected to be merely an exercise in the psychology of disputation. It should be clear that if they are to make any rightful claim upon the attention of logicians, fallacies cannot be simply false beliefs or false statements; they must have the syntactic form of an *argument*, i.e. a set of statements, one of which is a conclusion, the remainder the premisses [216, p. 3ff.]. Moreover, a fallacy is not just an invalid argument (shrewd De Morgan knew that there is no limit to these), but a kind of invalid argument that might well generally or systematically *seem* to be valid. It should not be required of an informal fallacy that it always seem valid (a) to everyone, or (b) to everyone whom it is directed, or (c) to everyone who advances it. A fallacy can easily be committed, either intentionally or not, by the arguer or by the 'arguee'. It should be required only that, to attain the status of fallacy, the logical blunder be a common snare and delusion of argumentation, exemplifying a broadly recurrent pattern of effectiveness, however, sophistical. Fallacy theory thus has a psychological aspect. It needs to be geared to errors of actual argumentation, and may be thought of as 'applied logic' or 'informal logic', in contradistinction to abstractly mathematical reconstructions of such classical notions as validity and implication.

The problem with most of the fallacies is that it is hard to see how the fallacy is an incorrect argument. The required concept of *argument*, being classically non-standard, requires analysis in an alternative and not well-defined formalism. Perhaps, however, the difficulty is not so pressing in the case of equivocation. For let F be any formalism adequate for the representation of a segment of a natural language. Whether such a formalism affords a philosophically worthwhile notion of argument is one thing; but equivocation is another matter, essentially a problem of translation, known traditionally as a semi-logical fallacy, a fallacy within language, or a verbal fallacy.

Quine suggests that the traditional fallacy of equivocation is a violation of a general principle of translation of sentences into the formal idiom: the trustworthiness of logical analysis and inference depends on our not giving one and the same expressions different interpretations in the course of reasoning [155, p. 48]. Thus the surface representation of the pair,

(1) He went to Pawcatuck and I went along.
(2) He went to Saugatuck, but I did not go along.

by the inconsistent schema $\ulcorner p \,\&\, q \,\&\, r \,\&\, \neg q \urcorner$ is equivocal. The 'I went along' in (1) means 'I went to Pawcatuck' whereas the 'I did not go along' in (2) means 'I did not go to Saugatuck'. Thus a correct representation of the pair (1), (2), is ⌜p & q & r & ¬s⌝ , a consistent schema. Quine adds that insofar as the background circumstances of an argument - such factors as speaker, hearer, scene, date, and underlying problem and purpose - remain constant, the fallacy of equivocation need not be feared. It is only when a contextual shift within the sequence of a given argument occurs that translation may be susceptible to the sophism.

It would appear that to understand equivocation is to understand the contextual shift in which one and the same sentence comes to represent a different statement than before. As such the problem is not so much proof-theoretic, and not so much semantic as pragmatic. It is useful to bear in mind that in a standard formal system where the domain is specified in advance, the possibility of equivocation is blocked. A *statement*, according to the prevalent understanding of this term, has exactly one signification, since otherwise it might be both true and false. In the usual interpreted formal languages, each sentence expresses exactly one statement, each name picks out exactly one element, and the risk of equivocation is nil. The accuracy of the translation into the formal mechanism forecloses upon equivocation. Nor can the related fallacies of amphiboly or accent be committed where the syntax is appropriately specified, the phonological component is suppressed. No, the medium of these three fallacies is the rich texture of natural language, where a phonetic string may admit of more than one syntactic representation, and where a syntactic string may admit of more than one semantical representation. We could sum this up as follows: if an argument is a set of what a logician would call statements, then no argument is equivocal. For a statement is univocal and does not admit of ambiguity.

2. CONTEXTUAL SHIFT

Let us turn to Quine's suggestion of contextual shift. Can this notion be made to help us understand what might be involved in mistaking a given argument for another, and distinctively different, argument? Not every ambiguity is an equivocation. For a fallacy, recall, is a fallacious *argument*. Quine claims that a kind of contextual shift in characteristics of the equivocal argument is involved. But why should such a shift occur? One primary explanation, or the key to it, may be found in a suggestion of Copi that the *plausibility* of the premisses (or conclusion) requires a contextual shift [40, p. 92f]. Consider the traditional example,

(3)   The end of a thing is its perfection.
      Death is the end of life.
      Therefore, death is the perfection of life.

According to the standard analysis of this venerable relic,[1] the term 'end' is ambiguous as between 'termination' and 'goal'. If the premisses are to be at all plausible, 'goal' is evidently meant by 'end' in the first premiss, and 'termination' in the second. On this view, we have before us four arguments, not one:

(4)   The goal of a thing is its perfection.
      Death is the goal of life.
      Therefore, death is the perfection of life.

(5)   The termination of a thing is its perfection.
      Death is the termination of life.
      Therefore, death is the perfection of life.

(6)   The goal of a thing is its perfection.
      Death is the termination of life.
      Therefore, death is the perfection of life.

(7)   The termination of a thing is its perfection.
      Death is the goal of life.
      Therefore, death is the perfection of life.

Now given the ambiguity of 'end', it is possible for a person to whom (3) is directed to interpret it as expressing any one of the four arguments, (4), (5), (6), (7). (4) and (5) are (plausibly) valid; (6) and (7) are invalid. Assuming that the person to whom (3) is directed seeks for a validity-preserving translation, he would choose (4) or (5). Now in (4), the second premiss is implausible, and in (5) the first premiss is implausible. True, in (6), both premisses are plausible, but (6) is invalid. So, like it or not, if the person to whom the argument is directed wants a validity-preserving construal, he is forced to decide between (6) and one of the pair, (4) and (5). What we have here is a classical case of cognitive dissonance [135], [62] - whichever way the 'subject' chooses, his target (correct argument) is denied him. He must choose between the invalidity of argument and the falsity of premiss, that is, between invalidity and unsoundness. One way of cancelling the dissonance is by an amalgamation of two arguments into one pseudo-argument having the appearance of validity and soundness both. Here then is the psychological explanation of the contextual shift. The victim of the fallacy chooses (6), foregoing the continuity of context demanded for validity, yet acquiring the illusion of it via the surface structure (3). Our victim sacrifices validity for premiss-plausibility, but maintains the illusion of validity nevertheless. (It is perhaps needless to add that interpretation (7) is rejected automatically, since it is not only not valid, but neither of its premisses is plausible.)

What makes the fallacy of equivocation work is the phenomenon of contextual shift. The person to whom the equivocal argument is directed is tugged to interpret some term one way in one premiss; yet another way in another premiss (or perhaps in the conclusion). But what produces the dual tug? Is it simply the all-important but ineffable factor of *context* or *background circumstances* of the argument? Yes, but more specifically, a major factor of context, a factor especially instrumental in giving rise to equivocation, has to do quite essentially with argument. An argument can only be deductively sound if the premisses are all true and entail the conclusion. If the argument is invalid, either a premiss is untrue or the entailment fails. The factor we speak of, then, is the tug jointly to satisfy these two conditions. We equivocate in our innocent attempt to make an argument sound. After all, arguments aspire after soundness; it is what they are meant to be.

Here then is at least an outline of one explanation of both the *apparentia* and *non-existentia* of equivocation. Nevertheless, harder questions remain, and we will now canvass a few of them. These problems also tend to have to do with pragmatic features of natural language, and to a striking extent, with special kinds of ambiguity that can there occur.

3. VARIETIES OF EQUIVOCATION

A main variety of equivocation shows itself in the use of the simile. Here is an example from William of Sherwood:

(8) Whatever smiles has a mouth
There are smiling meadows.
Therefore, there are meadows with mouths [210, p.136].

In our previous example, the word 'end' signified more than one thing *properly* (to use William's idiom), whereas in the above case, the word 'smile' signifies more than one thing *transumptively*, that is, metaphorically. In each case, the word is said to signify more than one thing on its own (*de se*) [210, p. 136], but in the present case we have equivocation *via* the use of a metaphor. Any theory of meaning adequate to the analysis of equivocation will therefore have to show itself ready to accommodate an account of metaphorical language.

Still another kind of equivocation arises in the kind of ambiguity in which one signification is central or basic, but where peripheral meanings come into play in special contexts; that is, where a word signifies more than one thing as a result of a special connection with something else rather than on its own (*de se*). In an example from De Morgan we have it that 'A person undertakes to cross a bridge in an incredibly short time: and redeems his pledge by crossing the bridge as one would cross a street, that is, by traversing the breadth" [50, p. 246]. This form of the fallacy, says De Morgan, arises from the mistaken assumption that the meaning of a compound expression is always a function of the meaning of its various meaningful elements. We might call this general kind of equivocation the *de aliud* equivocation, in the spirit of William of Sherwood.

Equivocations of this or a similar type sometimes have to do with the context of utterance or other special circumstances, such as tense-shifts. William cites the argument:

(9) Whoever was being cured is healthy.
     The sufferer was being cured.
     Therefore, the sufferer is healthy. [210, p. 136]

'The sufferer' normally means 'a person *now* suffering' relative to the time of discourse. Whereas in the second premiss, as a result of its connection with 'was being cured', it most plausibly refers to a time previous to the time of discourse, i.e. it most plausibly means 'a person *then* suffering'.

De Morgan's comments are insightful. In a natural language, idiomatic expressions are allowed. In certain idioms, the meaning of a single word is merged into the meaning of a sequence of words, the phrase or sentence in which it occurs, so that as it occurs in the idiom, its interpretation is tangential or peripheral to its 'original' meaning. What we have here is a kind of ambiguity but a special kind that is embedded in the pragmatic-contextual workings of the language. (Small wonder that equivocation is so often thought rightly to be an *informal fallacy*, a *fallacy within language*.) In seeking, therefore, for a theory of meaning adequate to the analysis of equivocation, pains need to be taken with idiom, with semantically 'non-functional' compound expressions, and with the contributions to such made by pragmatically contextual features of language.

Whately once said that sometimes equivocation is facilitated by grammatical transformations, for example, by paronymous words which do not always have a correspondent meaning, as in the following illustration [208, p. 195].

>   (10)   Schemers are not to be trusted.
> x has devised a scheme to prevent y.
>   Therefore, x is not to be trusted.[2]

This sophistry has given rise to the mistaken notion that everyone who devises a scheme is a schemer in the sense of that expression for which the first premiss is plausible. Here we are on the interface between ambiguity and amphiboly, i.e. *syntactic* ambiguity.

In all the cases we have so far examined, the equivocation comes from the ambiguity of a single term that occurs (twice) in the argument. However, there seems to us to be no particularly good reason not to allow that ambiguity of an entire premiss could also be made for an equivocation. That is, it would appear that sentential or syntactic ambiguity can also generate equivocations, as in this example:

>   (11)   Everything has a cause.
>   Therefore, there is some thing that causes everything.

Distracted by the equivocation, one might find this a convincing argument for the existence of One Cause. However, it is easy to see that the quantificational structure of (11) displays two arguments. Let a be a specific individual and 'C_{xy}' be a predicate for 'x causes y'.

>   (11a) $(\forall x)(\exists y)Cyx$
>   ―――――――――――
> $(\exists y)(\forall x)Cyx$
>
>   (11b) $(\forall x)Cax$
>   ―――――――――――
> $(\exists y)(\forall x)Cyx$

Here too the classical pattern of equivocation is evident. (11a) is invalid but the premiss is plausible. (11b) is valid but the premiss is implausible. The equivocator, in pursuing both the true premiss and valid argument, achieves a binocular blurring of (11a) and (11b) into the appearance of one argument, (11). Though the classical pattern of equivocation is clearly a factor in the mischief associated with (11), in this case too there is also an element of amphiboly. In fact, though in general equivocation is essentially a pragmatic affair having to do with contextual shifts, we can now begin to see that at times the motivation for such shifts is primarily semantic; as with (3) - 'the end of life'; and at other times the motivation is rather more syntactic, as with (11) - 'everything has a cause'; and at still other times the motivation is both semantic and syntactic, as with (10) - 'schemers', in which there are elements of both ambiguity and amphiboly.

4. GROSS AND SUBTLE EQUIVOCATIONS

For all their challenge to theory, the arguments we have examined so far are, in the main, simplistic text-book examples that would be unlikely to deceive a moderately attentive and perceptive audience. Indeed, it may be expected that the really interesting and dialectically effective cases of equivocation will occur in complex arguments where terms are hard to define and sometimes hard to grasp, and where the premisses may be widely enough scattered for the shift of meaning to pass unnoticed. Yet in just these cases it can be extremely difficult to support a charge of equivocation, for as Hamblin suggests, what is non-trivial may be controversial, and we may be hardpressed to show that a contested term really is univocal. [80, p. 15] Semantics is a slippery subject; intuitions about meaning are one thing; analysis are another. These problems too are legitimately a concern of applied logic and the study of the fallacies, even if, to the formal logician, they are 'merely' dialectical problems. Hamblin has pioneered a treatment of a number of these problems, and we turn now to a consideration of some questions that he has raised.

5. HAMBLIN: MORE PROBLEMS ABOUT MEANING

With at least two theses of Hamblin our disagreement has already been made quite apparent. One is the thesis that (i) equivocators must either deceive themselves or set out to deceive others into thinking their argument valid; the other is that (ii) since it is only *per stupiditatem* that anyone is ever deceived by gross equivocations, they are of little interest to the logician [80, p. 292]. Concerning (i), we have argued that although fallacies must generally be valid-seemers, whether or not an argument is fallacious or not does not depend on whether it seems valid or not to this or that particular person who may be the proposer or the receiver of the particular argument in question [229], [225], [222]. Concerning (ii), we have argued that certain gross equivocations may be, and need to be, used as 'text-book' examples to illustrate the logical form of the fallacy and serve as a model of the actual cases. Any broadly scientific enquiry may properly need to begin with a simple and somewhat idealized model which clearly exhibits certain germane and critical features of the case as a systematic first step toward a theoretical group of the subtleties and complexities of everyday cases. On the other hand, we do of course agree with Hamblin that the subtle cases disclose certain genuine difficulties of their own, to some of which we now turn [80, p. 292ff.].

(a) If someone is *deeply* deceived by what we consider to be an equivocal argument, there is the possibility that his own use of the sensitive words is not ambiguous. That is, his own usage may be 'non-standard'.

(b) Accordingly, it is sometimes difficult to know how to deal with cases of 'non-standard' usage, such as *stipulated* meaning. Certainly stipulated meaning can sometimes legitimately occur in argumentation, so it won't do merely to ban them.

(c) Likewise, it is in general difficult to see how a claim to 'standard' usage might be supported in the very kinds of cases that might most often figure in a charge of equivocation. Appeal to the intuition of the native speaker utilizing field methods of linguistics, for example, does not seem likely to issue in definitive, argument-settling results in just those cases of terms that often play a role in equivocation in argumentation. A list of certain terms that are especially and peculiarly liable to be used ambiguously, and an accompanying explanation of each term is helpfully provided by Whately in an appendix to *Elements of Logic*. [208, p.319ff.] The explanations run wildly to the philosophical, as might well be expected; 'can', 'may', 'must', 'argument', 'authority', 'cause', 'possible', 'law', 'experience', 'reason', 'sin', 'truth, 'value', 'wealth', and so on. 'Standard usage' is an elusive commodity in these regions, as those experienced in philosophical controversy will attest. Even if there were a 'standard usage' to be established, it is not clear that producing it would clearly settle whether the term was being used equivocally in a given argument, though, of course, it is always advisable for disputants to 'declare their stipulations'.

On the basis of considerations such as (a), (b), and (c), Hamblin suggests that the subtler variety of equivocation should sometimes be regarded less as a fallacy than as a disguised idiom. What seems to be an equivocation may simply be evidence of the creation of a new pattern of usage either established or developing in the language of the person who has advanced the argument, and for whom the argument is not at all equivocal; though, no doubt the balance would tip in favour of a verdict of equivocation should all stipulations have already been declared.

(d) What is needed is a view of equivocation that can allow for inventive or non-standard usage under certain conditions without allowing for total stipulative anarchy. Perhaps equivocation can best be seen as a procedural or dialectical concept that functions as a bar to systematic deviance from the usage set as standard relative to a given argument setting, for, as Hamblin quite correctly points out, [80, p. 294] we do not usually suppose a term or expression to be equivocal unless we get into trouble with it.

6. ARGUMENTATION IN NATURAL LANGUAGES

Equivocation is an affliction of argument within natural language. The ultimate goal of 'logical analysis' may be translation into a language with an exactly specified structure in which well-defined procedures on an antecedently specified domain regiment if not totally eliminate ambiguity. In practice, however, argumentation takes place in a less rarefied atmosphere, and guidelines for detection and avoidance of our fallacy must be applicable to actual practice. It is a characteristic feature of argument in natural language that it can tolerate ambiguity, and hence equivocation. But we propose that, even in practical logic, if we are to have what may be considered an *argument* in a natural language, certain boundaries must be set on what such a language may be said to comprise. For example, Tarski has shown that we must distinguish between expressions of the object-language and metalanguage if we are to avoid the possibility of inconsistency within the language. Hence any language suitable for argument should not be viewed as semantically closed [183], [182].

(I) The object-language of a disputation must be thought of as being at least roughly specified. That is, the participants must agree that the argument is to have taken place within some natural language (English, Swahili), or, a natural language enriched by a well-defined technical vocabulary (English + NASA jargon), or some quasi-artificial language, in which syntax and semantics must be at least roughly delimited. Even a purely stipulative language might be allowed, though, as with codes, there might be some reason to think that the notion of a natural language is being rather severely extended. And whether such 'languages' that are not codes could be supposed to be *private languages* is a question that we do not pursue here.

(II) It must be possible to specify the syntax and semantics of the language at least to the extent that a claim of ambiguity can be supported or arguably rejected. That is, there must be some *data* against which claims of ambiguity can be adjudicated [36]. There must be some form of data-processing procedure or set of agreements to handle questions of ambiguity independently of the mere say-so of the disputants on the question of equivocation.

(III) On the basis of (II), certain common ambiguities that occur in disputation might be standardly classified and recognized at the outset as ambiguous by participants in arguments. Whately's list is a helpful start.

(IV) In cases of argument within stipulated or artificial languages, there should be a syntactic structure rich and standard enough to support the generation of arguments in some suitable standard sense, certainly standard enough to justify the interest of the informal logician who seeks for an understanding of the fallacy of equivocation.

(V) Instances of stipulated meanings must be marked as such, and appropriate rules and definitions introduced by the proponent of the argument. Some metalinguistic conventions concerning the setting out of definitions, axioms, and the like, should be available. One example might be of procedural rules of dialectic of the type discussed by Hamblin is his systems of formal dialectic. These rules need not, however, be strictly formalized, for most purposes.

(VI) The analysis and discovery of equivocation would proceed by being clear about what is *stated* by the premisses and conclusion of an argument. As a beginning, therefore, premisses and conclusion should be clearly identifiable. Very often, of course, what is put forward as an argument is no argument at all, but rather a command, threat, promise, or some form of non-statemental utterance. To recognize equivocation, the student of practical logic need to appreciate and be able to apply the distinction between truth of a statement and correctness (validity) of an argument. As we have seen, the essence of equivocation is the blurring of this distinction in practice.

(VII) Metaphorical use of terms and the use of peripheral meanings, idioms, and analogies in argumentation are all areas where special caution is needed. In such contexts, equivocation should be looked for with special care.

(VIII) Peculiarities of syntax often contribute to serious equivocations, e.g. Mill's notorious pun on 'desirable'.

It is clear that we are here speaking as much of applied linguistics as formal logic. This may suggest to some that our restrictions are not really very important for the testing of arguments. To the contrary, we say. A charge of equivocation simply cannot adequately be dealt with in the absence of some such framework as is indicated by (I)-(VIII). Arguments most often are exercises within natural language, and it is in natural language where equivocation may occur. Exclusively formal methods of analysis and adjudication by themselves will very often be utterly unhelpful in providing a basis for judgments concerning equivocation. Our eight helpful hints depends for their practical usefulness on the development of an applied linguistics, and are, from the point of view of formal logic, mainly translation factors. Nevertheless, they are essential to a non-superficial theory of the fallacies with any chance of beginning to deal effectively with equivocations in actual argumentation.

7. CONCLUDING REMARKS

Equivocation may seem quite simple and theoretically untroublesome on the surface - it could simply be described as the appearance of validity conjured up[3] through the conflation of what are, in reality, two or more arguments. Yet the practical problem of the avoidance of equivocation in argumentation is far from simple. We have offered a set of working guidelines, but obviously these are barely the beginning. In fact, successful prosecution of the *charge* of equivocation is often exceedingly difficult to carry out against a determined opponent.

There is justification in the traditional practice of regarding equivocation as essentially linguistic, as a fallacy 'within' language. Much of the theoretical development of the notion of equivocation depends directly upon the analysis of the phenomenon of ambiguity in natural language. This is the task of linguistics, of the semantics of natural language. The eventual development of an effective working logic of equivocation must draw more of its inspiration from work in those quarters.

Chapter 16
Why is the Ad Populum a Fallacy?

The traditional informal fallacy of *argumentum ad populum* is standardly characterized as the fallacy committed by directing an emotional appeal to the feelings or enthusiasms of "the gallery" or "the people" to win assent to an argument not adequately supported by proper evidence. What is thereby characterized certainly finds the mark in pointing up a widespread ethical deficiency of advertising practices and a quotidian rhetorical shortcoming familiar in many aspects of public affairs. But is it a fallacy? Specifically what is wrong, as a deficiency of correct argument, with appealing to popular enthusiasm? And if this appeal can be a fallacy, exactly what manner of argument is it that is thought to be incorrect? In other words, exactly what is meant in this context by the phrase "not adequately supported by proper evidence"? These are hard questions, but they need to be asked.

We are told by a certain hamburger chain that not buying their product is virtually an affront to patriotic clean living, cheerful industrious dedication, and happy family togetherness. Wouldn't it be better if this commercial time were allocated to giving rational evidence that the food they sell is a good value or has arguable nutritional advantages? We are told by an oil company that nature is beautiful. Wouldn't it be better if this time were used to offer some rational assurance that this company is not destroying nature, or is at least contributing in some way to the quality of our lives? If so, then the question should be asked whether a fallacy has been committed, that is - whether there is an incorrect argument - or whether the deficiency is a breach of advertising ethics, or is some impropriety other than a failure of correct argument. But if the alleged inadequacy is in the argument itself, in its very logic, then our second question must be addressed. What precisely is the error? Is it an identifiable fallacy that admits of analysis and the determination of rational guidelines for adjudication?

These questions are worth asking not only because of the widespread seriousness of the *ad populum* and related sophistries as mischievous fallacies in practices of the manipulation of public opinion, but because there is a positive and constructive need to understand the fallacies, how they work, and when and why they are wrong. For a faith in the power of argument as an ally of truth and correct argument is the mainstay of our democratic political institutions and our adversary legal system. But in order to understand argument there is the scholarly requirement to bring some order to the study of the informal fallacies as a coherent discipline of logic and philosophy.

Students often perceive this need to understand the fallacies, but as Hamblin [80] has pointed out, the Standard Treatment largely fails to offer adequate theory to aid us in classifying putatively fallacious arguments into correct and incorrect cases. Too often, a putative fallacy like the *ad verecundiam* turns out to have instances that seem to be after all correct, even if the precise model of argument of which the cited example is a correct instance turns out to be elusive. Yet as John Woods and I have argued [225], [229], a fallacy, if it is to command our interest as an incorrect argument, must be more than a mere behavioral aberration, or a shortcoming of persuasiveness, manners or morals. It must be a wrong argument.

A main problem is that once it is perceived that the model of argument involved in an informal fallacy is not that of classical first-order logic, the move is either to dismiss the study of the fallacy as of no interest to the proper subject of logic, or to insist that "informality" must signal the complete inappropriateness of any structured decision mechanisms of any "formal" sort. Hence the stultifying bifurcation between formal logic and the study of informal fallacies that has resulted in the Standard Treatment.

In this essay I will try to indicate how that impasse should be overcome in the case of the *ad populum*, by arguing that formal mechanisms are involved, but that the logic involved is not standard, and that the role of extra-logical components must also be brought in and integrated with the formal elements. Thus it will be shown that what informality is involved need not require the entire avoidance of logical structures. Quite to the contrary. But the structures involved will not be those of classical logic.

1. ARGUMENTS DIRECTED TO SPECIFIC RECIPIENTS

Perhaps what initially seems most wrong or fallacious about the use of the *ad populum* is that such an argument is directed to a specific group of actual persons rather than being an attempt to argue from true premises. In arguing *ad populum*, when one selects premises, it matters little whether the premises are true. The question is rather one of whether these premises are plausible and will be accepted - if possible, enthusiastically - by the audience that is being confronted. The fallacy here would be that of throwing concern for the truth aside in favor of an outright partisan process of trying to convince by utilizing whatever assumptions, no matter how outrageous, that one's target audience seems prepared to tolerate. What seems wrong is that one's argument is allowed to be subjectively oriented, person-relative, and therefore it subverts the objective goal of arriving at the truth by the process of logical reasoning.

In this connection there seems a clear parallel between the *ad populum* and the *ad hominem* [223]. The main difference is that whereas the *ad populum* is directed toward the group, the *ad hominem* is directed toward one individual. Of course there is a difference or orientation in that the *ad hominem* is negative in its intent to discredit the individual, whereas the *ad populum* is positive in its intent to win the approval of the group. But the subjective element is common to both.

However, the above explanation of what is fallacious about the *ad populum* is incorrect, insofar as it rests on the assumption that the only worthwhile and legitimate function of argument is to reason from true premises - that is, to produce sound arguments. In my view, a properly broad and dialectical perspective on the concept of argument and its uses adequate to the study of the fallacies should include argument from premises that may or may not be true. Can we not, for example, argue from an opponent's premises, demonstrating that they imply a falsehood, in order to demonstrate to that person that one of his premises must be false? Do we not often, in numerous legitimate contexts of argumentation, work from premises of uncertain truth value in order to see if their logical consequences may better to enable us to assess their truth or falsity? If so, it is hard to see what is wrong in arguing from a set of assumptions that are accepted by an individual or group, but are not known by the arguer to be true. I am, of course, referring to the function of argument called by Aristotle *dialectical*. As the philosopher characterizes this notion at the very beginning of the *Prior Analytics*, a demonstrative premiss is one that is simply laid down by the demonstrator at the outset, whereas a dialectical premiss is one that the arguer's opponent is prepared to admit at the outset.[1]

What, then, is fallacious about a dialectical argument? Nothing, *per se*, we think, though of course dialectical argument, like any form of argument, is something that can be abused. So far, however, it is not clear how the *ad populum* can be located as a specific and identifiable abuse or fallacy.

Indeed, it is possible to see that there could be a fallacy in confusing or conflating dialectical with demonstrative argument.

If Bob thinks Sue is demonstrating, whereas in fact Sue is dialectically arguing from premisses well known to be false - despite their plausibility to poor Bob - then perhaps a fallacy of some sort is being perpetrated by Sue. But is it a fallacy of *ad populum*? Where is the mass appeal to the sentiment of the multitudes here? The fallacy is more one of equivocation. Sue and Bob have simply failed to clarify at the outset what the model of argument is that they are working with. The fallacy here is rather like that perpetrated by the Sophist who sets out to convince us that a good inductive argument is a bad argument on the ground that it is deductively invalid. If it is meant to be an inductive argument at the outset, and this is clear and agreed to by all participants, then the attempted refutation is beside the point. But surely it is not this sort of fallacy that is characteristic of the *ad populum*.

The fallacy we have identified is much closer to what Johnson and Blair [109, p. 158] call the *fallacy of popularity*, which occurs whenever an argument proceeds from the popularity of a view to its truth, viz.,

(A1) Everyone believes P.
      Therefore, P is true.

(A2) No one believes P.
      P is false.

As Johnson and Blair point out, (A1) and (A2) are deductive paradigms that, flatly stated, would fool hardly anyone. But they are forms of argument implicit in many a specious sequence of reasoning.

To sum up, it is sometimes perfectly logical or anyhow not fallacious to start with a dialectical premiss like 'Everyone believes that P' or 'This audience believes that P, so let us assume that P', but to make a move straightaway towards a conclusion of 'Therefore P is true' is illogical in general (exceptions where P is a statement like 'Someone believes that P'). (A1) and (A2) are by no means generally valid, and to advance them is to argue fallaciously.

However, (A1) and (A2) do not represent the entire fallacy of *ad populum* so much as what might be more accurately identified as the *argument from popularity*. Reason: there need be no emotional appeal to popular sentiments or mass enthusiasms unless P is a proposition of a special sort.[2] Nor need the premiss of (A1) or (A2) be irrelevant to its conclusion in any sense except that the arguments (A1) and (A2) are not universally valid.

In fact, (A1) and (A2) can be identified as failures of the respective inferences (I1) and (I2) below in a doxastic logic.

(I1) $(\forall x)$ (x believes that p)   (I2) $(\forall x)$ $\neg(x$ believes that p)
   To infer p.   To infer $\neg$p.

It would be reasonable to presume that neither (I1) nor (I2) would be rules of inference in any standard doxastic logic. Consequently, it is clear how (A1) and (A2) can quite legitimately be evaluated as fallacious moves in argument.

Yet it is often thought that the *ad populum* is essentially an emotional distraction by appeal to mass enthusiasms. True, (A1) and (A2) contain some element of "popular appeal" in the references to 'everyone' and 'no one'. But isn't there more to the *ad populum* than that? Our initial characterization of the Standard Treatment of the fallacy suggests that there is. Let us look to what else might be involved.

It is useful to note a possible exception of another sort to the invalidity of (A1) and (A2). If the 'everyone' refers to a collection of *bona fide* experts who are in a position to know about the truth value of P, then (A1) and (A2) may be legitimate arguments from expertise.[3] In this special case, we may be dealing with an *ad verecundiam*.

2. AD POPULUM AS AN EMOTIVE FALLACY

Ad populum, like its partners in crime, *ad misericordiam* and *ad baculum* [222] is often, perhaps usually, characterized as a fallacy that is essentially emotive. So construed, it is a questionable move because it attempts to short-circuit rational argument by jamming it with emotional interference. Why trouble to mount logical arguments if you can arouse your audience so much more directly and winningly by appealing directly to their raw emotions? The fallacy seems to consist in the element of mass emotional impact in certain appeals. As Copi [40, p. 79] puts it, "We may define the *argumentum ad populum* fallacy ... as the attempt to win popular assent to a conclusion by arousing the feelings and enthusiasms of the multitude". This way of characterizing the fallacy appears to be adequate to what the logic textbooks want to say about it, but is it a characterization that can be clearly understood as an identifiable logical fallacy?

First there is the problem that an appeal to mass emotion may be an attempt to waive or subvert argument, but is may not itself be an argument. Where are the premisses and conclusions to be found? And indeed, it is by no means clear at all what is wrong, logically or otherwise, with attempting to win popular assent to a conclusion by arousing the enthusiasm of the multitude. Surely if the conclusion is worthwhile, such an attempt could be, and often is, highly commendable.

But the overwhelming difficulty with this approach is that it would seem scarcely credible that there could be decision procedures or even rational guidelines for determining that an argument is incorrect exactly when an attempt has been made to win popular assent by arousing mass feelings. The fact is that any attempt to define the fallacy in this way would be enmeshed in a hopeless and vitiating psychologism of the worst sort. Once we start defining incorrectness of argument in terms of attempts at arousal of mass enthusiasms, we can be sure that the basic problem of identifying the species of incorrect argument that constitutes the conceptual core of the fallacy has become unmanageable. We have pretty well exclusively gone over into the psychology or sociology of rhetoric and propaganda. Consequently we are back to the first problem -where there is no argument, there is no fallacy, because there is no possibility of the notions of correct or incorrect argument being applicable.

Whether or not mass feelings are aroused should not be a determining factor in evaluating the correctness of an argument. The mass enthusiasms are surely not an element of the argument or a property that the argument has. Rather they are a property of the audience, the target group to whom the argument is directed.

But the final difficulty with the approach under consideration is the plain fact that perfectly correct arguments can appeal to mass enthusiasm. Enthusiasm is not fallacious.

3. IS IT AN ARGUMENT AT ALL?

We have argued in [233], [229], [223] that to be fallacious an argument should be incorrect or invalid in some clearly specifiable way. But at very least, we would maintain, a fallacy should be a fallacious *argument*. What exactly an argument is poses a question of the basic import for serious students of the fallacies, needless to say: [80], [233]. But whatever an argument is, a quite generally applicable minimal necessary condition is that an argument should be a set of statements (propositions), namely the premisses and a conclusion. Erotetic and other evidently non-statemental fallacies may possibly be exceptions - and then again they may not - but these issues aside, it is at least strongly presumptive that anything that cannot be viewed as a set of statements may be questioned as to whether it can constitute an argument that is definitely a worthwhile object of study in the domain of the logic of the fallacies.

The *ad baculum* is a case in point - if I threaten you, how if at all do I argue? How then do I argue fallaciously? In our article "*Ad Baculum*", [222] we fail to find a concept of argument that fits the textbook model of the *ad baculum*, and so query whether the *ad baculum* can ultimately maintain its reputation as a fallacy. The very same difficulty is inherent in the *ad populum*. Does the demagogue even argue when he whips his audience into mass approval by flagellating powerful emotions? May he not rather be avoiding argument altogether? It is a good test to see if premises and a conclusion may be found.

A well-known lumber company makes successful use of the popularity of the "back-to-the-land" and "do-things-for-yourself" North American folklore, especially popular in the counterculture of recent years, to appeal to a wide stratum in its population of possible buyers. Rather than speak of the quality of the tools, lumber, or other products it advertises, their commercial message conveys feelings of accomplishment and pride that are associated with building things for yourself. A point has been evaded, true, but the advertisement hits a popular target by appealing to the pride of personal do-it-yourself accomplishment. Perhaps then, something evasive or manipulative has transpired. But where is the fallacy? Indeed, where is the argument? The fact is that a large part of what seems to be wrong is that argument of any sort has been foregone in favor of a direct and quite successful appeal to widespread sympathies and attitudes. Where are the premisses and conclusions in such an appeal?

Yet is may be felt that the fallacy consists in the evasion itself. What is wrong is that the emotional appeal is somehow irrelevant, a deception by distraction. Indeed, many textbooks propose that what is fallacious about the *ad populum*, *ad hominem*, and other fallacies associated with emotional appeals is that the emotional appeal is *irrelevant* to the conclusion of the argument. What remains to be seen, however, is how the requisite notion of failure of relevance interacts with there being an argument, and what sort of relevance is involved. What we are being told is that in effect the *ad populum* is a species of *ignoratio elenchi*, or misconception of refutation. Our lumber company has committed a fallacy by simply changing the subject from the topic of the quality of their products to the topic of the popular folklore of back-to-the-land fantasy.

The suggestion may be that we should be working in a relevance logic or relatedness logic. It has been shown in Walton [200] how relatedness logic is applicable to the *ignoratio elenchi*, so let us look to treating *ad populum* in a similar way. P implies

Q in relatedness logic if, and only if, it is not the case that P is true and Q is false, and there is subject-matter overlap between P and Q.[4] To determine if there is subject-matter overlap take a set T of the most specific possible topics. Then assign to P and Q the subject-matter *p* and *q* respectively, where each subject-matter is a subset of T. P is said to be related to Q in this sense if there is at least one topic in T that is in both *p* and *q*.

True conditionals in classical PC like 'The moon is made of green cheese, therefore 3 is the square root of 9' fail to come out true in relatedness logic provided the two putative implicationally related propositions fail to admit of common subject-matters. 'Not-p implies p-implies-q' is not a theorem[5] and therefore is a formal fallacy, from the viewpoint of relatedness implication, when connectedness of subject-matters is thought to be a part of the argument. Of course there is nothing wrong with this theorem if connectedness of subject-matters is not at issue. In that case, classical logic is perfectly applicable. Indeed, we may say that classical logic represents the philosophical view that all propositions really are connected by subject-matter overlap, and therefore in classical logic matters of connectedness in topics need not be specified or taken into account. On the classical approach, 'Not-p implies p-implies-q' is acceptable because the assumption is that p and q really are related in some fashion.

If topic-sensitivity is required, and the requirement of implication is that there be subject-matter overlap between implicans and implicandum, then relatedness logic is clearly more applicable than classical logic or other alternatives. Thus if the failure of argument is one of topic relevance, the use of relatedness logic would seem to be a primary candidate for analysis of the model of correct argument concerned.

We suggest that there is considerable promise in this line of approach to the *ad populum* and also the *ad hominem*, provided a relatedness logic is brought into play in order to define clearly the semantics of subject-matters in implications. But it is an approach that needs to be pursued with some circumspection. There are several grounds for caution.

First, it is clear that failure of relatedness is at best a necessary condition of the *ad populum*, for one's premisses may be unrelated to one's conclusion without there being any element of mass appeal characteristic of the *ad populum*. That is, *ad populum* may be one species of *ignoratio elenchi*, but the question remains unanswered of how it can be identified as that species.

Second, it is not clear that failure of relatedness is even a necessary condition of the *ad populum* because in the most outrageous *ad populum* there might be considerable subject-matter overlap between the conclusion and the statements that form the basis of the mass appeal.

As Ellul [56, p. 84] points out, genuine information can often be mixed in with propaganda in order to heighten the overall effect of the propaganda. In advertisements for automobiles or electrical appliances there are often legitimate facts about technical specifications or proved performances mixed in with the appeal to feelings and passions. Thus, even in an argument that is an *ad populum*, there might be considerable subject-matter overlap between the conclusion and some premises.

If subject matter overlap occurs, still the *ad populum* may fail as a relatedness implication because it is not true that the premisses imply the conclusion in virtue of the truth values.[6] In other words, sometimes an *ad populum* may be thought to be an "irrelevant appeal" simply because the conclusions can be false even if the premises are true. This may explain why sometimes an *ad populum* or other fallacy of failure of relevance may seem to have occurred even if relatedness exists in the argument.

To sum up, relatedness implication is at best a necessary condition for the analysis of the *ad populum* and even so, still does not define the fallacy or differentiate it from the *ad hominem*, *ignoratio elenchi*, or other fallacies that arise through failure of relatedness.

The third difficulty is the one that has persisted all along, namely that it is hard to see very often how what is characteristically cited as an *ad populum* can be construed as an argument, if only because the statement requirement seems too often not to be clearly met. Copi [40, p. 80] exemplifies the standard conception of the *ad populum* by claiming that it is effectively used by advertisers who glamorize their products by sketching for us daydreams and fantasies that excite the approval and admiration of the average consumer. A fantasy may make a statement of sorts, but can we assume that a fantasy is a set of statements with determinate truth values? We can't help having some reluctance in acquiescing too readily in such an assumption.

4. IS IT A SINGLE FALLACY?

We can summarize our findings in the observation that the standard conception of the *ad populum* is a composite of four elements.

1. A doxastically invalid inference from a set of beliefs by a group to a conclusion of impersonal

truth. This could be called the fallacy of popular appeal.
2. The deviation from argument. What is involved here is not incorrect argument but simply the abandonment of argument altogether.
3. Failure of relatedness. The fallacy here is that of *ignoratio elenchi* - incorrect argument by way of disjoint subject-matters.
4. The element of mass enthusiasm. The appeal to "the people" or "the gallery" is an emotional appeal designed to rouse the feelings and enthusiasm of the multitude.

We are suggesting that these four elements can be clearly distinguished from one another even though they are related in the practice of *ad populum* arguments. For example 2. is presumably made as a move in order to utilize 4. To see how they are connected observe that 4. is used as a distraction to obscure the fact that 2. has occurred. 3. occurs by way of 4., and is presumably often successful because of the distracting nature of 4. Moreover, clearly 1. and 4. are connected - both are an appeal to the group of which the target audience is a subset.

But there are differences of significant methodological import. 1. and 2. both pertain to classes of deductively invalid inference forms, but the deductive system is by no means the same in each case. 1. pertains to doxastic logic, and 3. pertains to the semantic interpretation of relatedness logic. 2. is quite a different matter from either 1. or 3. 2. is not a question of invalid argument, but is rather a failure to meet the conditions of being an argument. Some would say that 2. is a matter of logic, others would not. It depends on whether logic is said to include the identification or location of arguments as well as the task of their evaluation into the classes "valid" and "invalid". On more liberal views of the scope of logic, or perhaps better, the analysis of argument, 2. might count as a task of argument analysis for which there should be a place in applied logic. But 4., by contrast, is a frankly empirical element. Whether or not there is an appeal to mass enthusiasms or popular feelings is surely a matter to be decided and evaluated by sociology, psychology, or rhetoric as empirical disciplines. Thus 4. stands out as an element that is distinctive and somewhat out of place with the others, theoretically speaking.

The lessons we have learned to this point are largely negative. None of 1. to 4. taken singly can represent adequately what is standardly taken to be the *ad populum* fallacy. As for 4., we saw that there is nothing wrong with mass enthusiasms. Turning to 3., we recall that failure of relatedness is at best a necessary condition of the *ad populum* fallacy. As for 2., we argued that if there is no argument, and therefore no incorrect argument, it is hard to see how there could be a fallacy. 1. is certainly a fallacy of sorts, but does not seem to represent the full measure of the *ad populum* unless the other three elements are also present. Can these elements be somehow fitted together, to yield a clear, coherent, adequate and workable analysis of the *ad populum*?

The key to understanding how these elements should be integrated is the distinction between two properties of argument - correctness and effectiveness. The logical core of a fallacy lies in the normative question of whether a specific instance of the argument is correct or not. The task here is to set out decision procedures for evaluation of argument by constructing an abstract model that meets theoretical requirements of clarity and precision. If possible, the model should be formal.[7] If not, it should at least exhibit some theoretically well-defined structure, however tentative or unrefined. For otherwise, in the opinion of the present author, clear and objective guidelines will not become available.

Whether an argument is correct or not is, however quite a separate question - in a universe of non-omniscient arguers and recipients of arguments - from whether the argument is effective in the sense of its capability to modify the credibility of a target audience in the direction desired by its proponent. The latter aspect was called the *apparentia* (seeming-correctness) of on argument by William of Sherwood and other medievals, as opposed to its validity or correctness.[8]

Whether an argument is effective is essentially a psychological or sociological matter, unlike the question of correctness, which is a matter of logic. Psychology and logic are not entirely unconnected - despite the well-known danger of an unfashionable psychologism - for the study of the fallacies and their applied logic is just one major area where they do intersect. But in principle, the two matters of correctness and effectiveness must be clearly distinguished so that their interrelationship can be carefully studied and not confused. For the danger of confusion is precisely that of making the fallacies inaccessible to study, after the fashion of the Standard Treatment.

We can see now that 4. represents the psychological vehicle of the *ad populum*, the *modus operandi* of its *apparentia*. True, there is nothing wrong with mass enthusiasms in themselves, but they become dangerous when they make possible the plausibility and attractiveness to an audience of any one of the fallacious moves of 1., 2., or 3.

One thing that can go wrong in an *ad populum* is that emotional turbulence permits a failure of relatedness to go without censure or detection. The argument strays off topic but the emotive distraction offered by 4. obscures the failures of relatedness.

Yet even more drastically, the emotional element can become so dominant that a complete abandonment of argument takes place. Not only does relatedness fail between P the premises and Q the conclusion, but the statement requirement itself even fails to be met, so that what we have is not a proposition P or Q at all but a mere semblance thereof. In a sense therefore, as a failure of argument, 2. is more severe than 3. Yet in either case, the move is made plausible through the psychological matrix of 4.

Can we now have exhausted the full import of the *ad populum*? 3. by itself is not both necessary and sufficient for the *ad populum*. Nor is 2., because 2. is marginal as a full-fledged fallacy – it represents not an incorrect argument but a lack of argument. Similarly, 4. is not an essential characteristic of the incorrectness of the *ad populum* but is rather its psychological *modus operandi*. But put all three together. Here we have a fair account of the fallacy: it is a deviation from argument or a failure of subject-matter relatedness in argument, in either case made plausible by an effective appeal to mass enthusiasm.

But this account of the *ad populum* is not quite good enough, because the fallacy so defined turns essentially on the empirical question of what constitutes a mass enthusiasm. We must remember that 4. is a property related to effectiveness, yet the standard conception of the *ad populum* seems to require that the element of popular appeal is an essential part of the logical core of the fallacy. If so, then 1. must be brought in as a deficiency of argument that pertains specifically to the appeal to a group of persons that is characteristic of the inner logic of the *ad populum*.

Weighing the value of all four elements in attaining the best possible overview of the *ad populum*, we propose that we take the fallacy to consist in a disjunction of 1., 2., and 3. That triad explains how it can be a deficiency in argument or an incorrect argument in one of three characteristic ways. But then 4. shows how these three failures – unimpressive moves of argument as they seem, exposed in their bare essentials of incorrectness – become a powerful strategy of deceit and mischief when employed through their *modus operandi*, 4. It is 4. that assures the practical importance of the *ad populum* as a significant error of argumentation. But it is the disjunction of 1., 2., and 3. that explains what is wrong.

Chapter 17
What is Informal Logic?

Theories benefit from formal methods in well-known ways. Such methods can yield up perspicuous structural representations; they can furnish taxonomic and definitional depth and clarity; they sometimes can provide for the effective or anyhow non-constructively successful recognition of a theory's target-properties. Perhaps it may be assumed that a body of knowledge is non-trivially eligible for formal treatment when (1) the objects of theory enter into interesting systematic interconnections expressible in functional or quasi-functional ways; and (2) such interconnections obtain or not, as the case may be, under semantic suppression of the connected items. Functional connectedness amidst semantic suppression may, then, reasonably be said to be the essential basis of the fruitful deployment of formal methods. But there may be other features of a theory's domain that bid welcome to formal techniques. In logic, for example, we deal with infinite domains of sentences (or, if you like, propositions or thoughts) and with unbounded sentential and argumental operations upon them. The law-like interconnections that logicians seek to record in the appropriate metatheorems are general in their sweep; and generality over infinite subject matters is known to be clearly and manageably expressible in formal terms. Indeed, to such *philosophical* questions as, "How, with only finite attention and finite resources, can such infinite generalities be wrought or thought?", the formal logician can sometimes proffer the notion of finite axiomatizability within complete, consistent and sound logistic systems.

1. APPLICABILITY OF FORMAL METHODS

Suppose now that someone identifies "The cow kicked the mule, hence the cow kicked the mule" as a circular inference. He might go on to add, quite correctly, that so too is any inference similar in salient respects. While perfectly true, this is an oblique generalization; its infinite upshot is only alluded to, not identified.

The formal logician has a better way of saying much the same thing: "Any inference, [A, hence A] is circular." It is better because the use of variables accommodates infinite upshot with finite resources, but also because, for any object you like, it is infallibly and mechanically ascertainable in at most finite time whether or not it is an inference, [A, hence A.] So we have recursive enumerability and decidability.

A further advantage of this sort of formal approach is that it demonstrates that, and the extent to which, possession or lack of target-properties (e.g. circularity) is not a matter of parochial semantic status, and not a matter either of parochial contextual and pragmatic features. Now, to be sure, beyond the frontiers of simple first-order theories, target-properties can be expected to involve semantic, pragmatic and contextual considerations that first-order theories need not and cannot represent. But the fact remains, if such non-first-order theories are to be formal theories, there will always be parochial semantic, pragmatic and contextual irrelevancies, and they will always be suppressed.

When you have formal representability, recursive enumerability and decidability over infinite domains of semantically suppressed primitive items systematically dispersed in functional ways, then you have, or are close to having, an abstractly mathematical system, M. The system M may *itself* be a good candidate to be an object of study, an object of foundational or metamathematical scrutiny, itself organized formally and axiomatically. If the theory that M actually expresses is a logical theory - for example, a theory of argument-circularity - it may also be true that, considered as an object of foundational attention, M is metamathematically interesting in ways that have little to do with the fallacies. It is useful to keep clearly in mind that the original motivation and the metamathematical motivation may be quite different, and that they can involve M in essentially different ways, and that some of what your metamathematical investigations may disclose about M will do nothing to illuminate the *petitio*. Of other such lessons, however, we are not so sure. Would it, for example, be helpful for a clear understanding of the *petitio* that M be metamathematically determined to be isomorphic with Kripke's intuitionistic semantics or with Hintikka's new system of dialogic?[1] Quite possibly so. In any event, should M be a fit object of metamathematical attention, two kinds of illumination may be forthcoming in the metamathematics. (1) There may be a richer appreciation of M's *subject matter*, and (2) one may learn something of M's own expressive and demonstrative limitations and capacities.

Plainly, then, it would seem that being a mathematical system is not necessarily a liability for a theory of the fallacies, but, on the other hand, it certainly is not necessary that fallacy-theory be a mathematical system if the fallacy-theory is to be instructive or even deeply correct.

One of the most persistent prejudices against the use of formal methods in the treatment of the fallacies comes from confusing the use of formal methods as such with *formalization*, that is, with the construction of what Church calls logistic systems. Now it is quite true that a clear understanding of a logistic system is no guarantee of a clear understanding of the body of doctrine that it formalizes. At times a formal definition will be little more than "a platitude restated in pedantic obscurity". (Consider, for example, the usually formal definition of the continuity of a curve.)[2] Then, too, there are important limitations in relation to motivating intuitive concepts. We cannot axiomatize without residue either the intuitive notion of positive integer or the basic notion of set.[3] It would surprise us greatly if consistent, complete and sound logistic systems were found to be intuitively adequate to such notions as *argument*, *part-whole*, *expertise* and *burden of proof*, all of which, after all, are needed for fallacy-theory. But what are we to take from this? I believe that we may say that if formal treatment involved nothing but the construction of axiomatic logistic systems, then the theory of the fallacies almost certainly could not in any non-trivial sense be a formal theory. We know, however, that axiomatic formalization does not exhaust formal treatment; and so the prejudice against the formal pursuit of the fallacies requires a different justification.

It is, of course, quite obvious that some bodies of doctrine do not formalize deeply, extensively or at all. This may indicate that certain objects and their target-properties are by nature insusceptible of all but trivial formal management. It may *also* indicate merely that extensive use of formal methods is premature, that the intuitive terrain is taxonomically too unsettled and definitionally too unbounderied, and its basic conceptual geography too little known to admit of formal reorganization. We have no hesitation in saying that much of fallacy-theory fails of formal treatment because the data to be formalized are not intuitively well-enough understood, are taxonomically and definitionally unfocused, and so on. But we cannot bring ourselves to a more austere skepticism than this. Of the dozen or so fallacies that we have been studying recently[4] there is not one case in which the investigation did not benefit from the application of formal methods. Graph theory and intuitionistic logic are, we think, helpful in modelling circularity;[5] causal logic fixed perspectives for the *post hoc*;[6] Hintikka's system of dialogic gives an interesting representation of dialectical exchange; Routley's consistent and complete system of dialectic illuminates certain features of the *ad hominem*;[7] various constructions of erotetic logic work well for Many Questions;[8] and so on. In our own work, we have been impressed to discover two particular advantages in the deployment of formal resources. One is the provision of clarity and power of representation and definition. The other is provision of clarity and power of representation and definition. The other is provision of verification *milieux* for contested claims about various fallacies. As for representational or definitional headway, we repeat that circularity models well in Kripke's intuitionistic semantics, and that a reasonable notion of evidential cumulativeness is also there definable [226, p. 23-5]. Then, too, Burge's formal theory of aggregates furnishes one with a quite powerful (though not effective) command of part-whole relations, and the theory of composition and division plainly benefits from this [28], [228]. (By way of illustration, we develop this last point in the Appendix.) Berger's connectibility logic provides a notion of causal connectibility defined over a four-dimensional differentiable manifold. It has occurred to us that this definition allows one to represent the *post hoc* as a modal fallacy of arguing from possibility to actuality [20], [236, p. 581].

We said that formal systems may also serve as verification (and for that matter, falsification) *milieux* for disputed claims about the fallacies. For example, if one places inference-structures (as opposed to entailment-structures) in a classical logic, then Mill's contention that the syllogism is a circular form of inference seems next-to-confirmed.[9] But if your base logic is relevant logic, it is next-to-disconfirmed. The usefulness of this observation may consist in its suggestions that relevant logic is a better (perhaps only marginally better) logic of inference than classical logic, and perhaps, too, that classical logic is a better logic of entailment than relevant logic. Or, to take another example, if you reconstruct the part-whole relation mereologically, then no inference from part to whole is a fallacy, and a great many inferences from whole to part which are not fallacious would be classified as fallacies nevertheless.[10] From which it may be concluded that the salient part-whole relation is not mereological (any more than it is set-theoretic).

2. CONTEXTUAL PECULARITIES

In addition to the formal devices that we have already spoken of, we have found it necessary variously to repose the theoretical burdens of the fallacies in probability theory, acceptance theory, epistemic and doxastic logic, and rationality theory.[11] Few of these systems are fully axiomatic; fewer still offer much promise of recursive enumerability and decidability with respect to basic intuitive notions or target-properties. Each is, in its own right, an object of some degree or other of controversy and uncertainty. None is a perfect instrument of analysis. At this stage of their theoretical development, the fallacies do not admit of and do not need grand axiomatic reorganization; for one thing, you cannot reorganize what is not yet organized. But this not more counsels against formal techniques for the fallacies than the open-endedness and non-effectiveness of aggregate theory counsels against Burge's impressive formal treatment of that subject.

Speaking of aggregate theory, it is useful to point out that it is an empirical theory, not a logical one,[12] and it is probably true to say that rationality theory, of the sort the Kyburg, Harper, van Fraassen and others are working on, will develop more as a contributions to epistemology than to logic. Then, too, it hardly seems correct to associate dialogical systems such as those of Hamblin and Hintikka less with games theory than with logic. *This leads us to suggest not that the mature theory of the fallacies is a branch of logic that is essentially informal, but rather that the mature story of the fallacies is a branch of formal theory that is essentially extralogical in major respects.* The formal theory of the fallacies is not (just) logic. What else it is or may prove itself to be remains to be seen; but it is vastly unlikely that it will not involve those extralogical theories that we have just met with.

More than one of my colleagues has met this suggestion with a certain civilized horror. Is not rationality theory at least in part a psychological venture? Indeed it is, and the authors have (brazenly or inadvertently, who's to know) saddled themselves with a gross and ridiculous Psychologism. Did the great Frege rail against his heresy to not avail?

Psychologism. Let us identify a rather silly kind of blunder. One commits it by reasoning as follows: If T is a theory with domain D and if $\alpha_1...\alpha_n \; \varepsilon \; D$ and are (or are representations of) mental contents, and if the consequences of T include sentences in the form; [A(... α_2...)], then such are psychological sentences and T is given over to Psychologism. Perhaps charity would recommend that this appalling argument not actually be attributed to anyone, but accuracy commands that mention be made of Carnap's criticism of Russell's in "Empiricism, Semantics and Ontology". It will be recalled that Carnap was complaining of Russell's "Psychologistic" view that propositions are mental events.

Historically speaking, a theory was Psychologistic when three conditions were met:
(i) The truth of the consequences of T was subjectively determinable (or some such thing).
(ii) This was so because such propositions were implicitly about one's own states of consciousness to which one alone had reliable access.
(iii) The consequences of T included ascriptions of certain target-properties that were *normative* (or some such thing); for example, validity, correctness or consistency.

Is this, then, what an interest in rationality theory commits us to? We don't see that it is; we doubt that it is; but we are prepared to wait and see.

We have been saying something about why I think that the fallacies are usefully pursued within formal theory. On the evidence, the fallacies involve systematic interconnections unvarying among items that themselves admit of some degree of parochial semantic, pragmatic or contextual variation. Thus the domain of theory can reasonably be taken to be a infinitude of abstract (and somewhat idealized) object interacting in somewhat functional ways. Such a set-up *suits* even if it does not call out for formal address.

Of course it is true that all our actual arguments are contextually unique, and it is possible, I suppose, that their formal construal will leave un- or under-represented those properties that make for such uniqueness, and moreover that these are the crucial properties for determining the adequacy or otherwise of those actual, real-life arguments. But if contextual or semantic peculiarities always carried the evaluative day, one could bid farewell to fallacy-theory, never mind whether or not formally transacted. And if contextual or semantic peculiarities always carried the evaluative day, one could bid farewell to fallacy-theory, never mind whether or not formally transacted. And if contextual or semantic peculiarities always carried the day, then farewell, too, to first order logic. Unless and until it can be shown that an argument's status regarding e.g. circularity is inextricably bound up with its individual peculiarities, whereas its status regarding e.g. deductive validity escapes such idiosyncratic particularities, then the undoubted contextual singularity of real-life arguments no more dismisses the one preoccupation than the other from serious theoretical prospect. (No more than would the undeniable contextual peculiarities of the dome of St. Peter's basilica prohibit the attention of theoretical metallurgy while welcoming the attention of theoretical physics.)

The contextual peculiarities of real-life arguments make for a point of some importance nevertheless. The intuitive notions of argument, inference and the like ar not (so far as we can believe) recursively enumerable, to say nothing of recursive. And it is arguable that the connectives of truth-functional logic are, each of them, less than wholly adequate reconstructions of their counterpart conjunctions in English. Thus we have it that the objects of truth-functional logic are not, strictly speaking, real-life arguments made up of real-life statements, involving the operation of real-life conjunctions. No, the objects of truth-functional logic are formal objects which, in one or other allowable way, represent the real thing. It is always nice to question as to the success or failure of such a formal theory's fit with reality, that it to say, as to the success or failure of its *application*. The same is true of any formal theory of the fallacies; it may always be open to question whether any such theory has a wholly satisfying application in the realm of actual argumentation. But that no more disqualifies the formal theory of the fallacies than it does truth-functional logic.

It is worth repeating that almost certainly axiomatization is not yet (if ever it would be) the way for fallacy-theory to develop.[13] This seems to us not to be the time for logistic systems. But neither it is sound to commit the story of the fallacies to what we might call "situational logic", that is, to an enterprise of merely anecdotal casuistry which over-concentrates upon the non-recurring case in all its ineffable singularity. What the fallacies now need is not the radical Fox but the moderate and common-sensical Hedge-hog.

3. FALLACIES

We have been assuming throughout that the principal content of what is so often called "informal logic" is the fallacies - *accentus*, affirmation of the consequent, ambiguity, amphiboly, *argumentum ad baculum, argumentum ad hominem, argumentum ad ignorantiam, argumentum ad misericordiam, argumentum ad populum, argumentum ad verecundiam*, begging the question, composition, denial of the antecedent, division, equivocation, *ignoratio elenchi*, illicit process, many questions, *non causa pro causa, non sequitur, post hoc, ergo propter hoc, quaternio terminorum, secundum quid*,[14] as well as the several fallacies that Alex Michalos has invented and the few that he commits; and of course a theory of argument that is sensitive to all this complexity. If this has been a tolerable assumption, then we have an answer to the question with which we began, "What is Informal Logic?" *Nothing is*. The theory of the fallacies is not logic, though it includes some logic, indeed quite a bit of logic; and the theory of the fallacies is not only at its best as a formal theory, it is difficult to see how the suppression of its formal character could leave a residue fully deserving of the name of theory.

Now, this is not to deny that, on a quite different interpretation of "informal", there do exist perfectly legitimate and familiar instances of informal "logic". An analogy with mathematics might serve the point at hand. Mathematics that is done in the usual, workaday way, that is to say, in ordinary mathematical English and prior to any axiomatic treatment, is said to be informal mathematics. There is no reason to deny to fallacy-theory this same kind of informality. In both kinds of case, informality is a pre-axiomatic affair, and we have been at some pains to persuade the reader that the construction of logistic systems is not by any means the only, or best, way to employ formal methods.

And, of course, further semantic confusions still are possible. One might take a fallacy-theory to be informal just because it is not worked up within a strictly *deductive* framework - or, for that matter, a framework of *inductive* logic. (Nor should it be.) Or, one might suppose a treatment of the fallacies to be informal if it stresses the complexities of the *application* of theory to the on-going scene. Further still, the fallacies might commend themselves to our attention from the point of view of *praxis* - as manuals of self-help for the ratiocinatively insecure.

But note. *These* enterprises do not preclude the quite vigorous exercise of what we have been calling formal methods. On the contrary, they very much *require* such an exercise if they are to attain the generality or power that commands serious philosophical attention.

4. ILLUSTRATION OF USING FORMAL THEORIES: THE FALLACY OF COMPOSITION AND DIVISION

The fallacies of Composition and Division involve misinferences from parts to wholes and from wholes to parts. An account of such fallacies must contain an account of relations of part to whole.

1. *Are Wholes and their Parts the Same as Sets and Their Elements?*

The short and easy answer is No. Sets are not spatio-temporal objects and do not admit of being acted upon or perceived. Sets are abstract; and where their members are concrete, most interesting composition and division questions are answered automatically (and, too often, erroneously) in the negative. Sets and their members do not simulate a type of whole-part set-up adequate to the analysis of composition and division.

2. *Are Wholes and Parts the Same as Suppesian Bodies and their Parts?*
A better choice than set theory for the reconstruction of the part-whole relation would be something like mereology in the fashion of Lésniewski [123]. For one thing, a significant aspect of the motivation of mereology was to hit upon a theory of classes, not only as actual collections of their objects, but also as collectivities which would mirror some of the features of collective as opposed to distributive predication. An interesting development of what is basically a mereological theory can be found in Noll's contribution to a theory of bodies adequate for the foundation of mechanics, and in an extended and refined version of Noll's account, due to Suppes, a version which we shall follow here.[15] One needs a workable notion of part in order to plumb the "logical" structure of composition and division. Bodies, paradigmatically perhaps, have parts, and so one may expect a theory of bodies to say something useful about parts.

In the Noll-Suppes theory 'π' can be taken to designate the part relation. The following definitions ensue:

If B π A and C π A then A is an *envelope* of the pair-set $\{B, C\}$. A is the *least envelope* of $\{B, C\}$ if, and only if, A is an envelope of $\{B, C\}$ and for any X that is an envelope of $\{B, C\}$, A π X.

A is a *common part* of $\{B, C\}$ if, and only if, A π B and A π C. If A and B are bodies, they are *separate*, if, and only if, they lack a common part. If A and B are bodies, then A is a least part of B if, and only if, A π B and there is no body C such that C π A and C = A. Body A is the *greatest common part* of $\{B, C\}$ if, and only if, A is a common part of $\{B, C\}$ and for every body X, if X is a common part of $\{B, C\}$, X π A.

Furthermore, if A is the *least envelope* of $\{B, C\}$, then the set-theoretic union of B and C is identical to A; A, then, is the *join* of B and C. If A is the greatest common part of $\{B, C\}$, then the set-theoretic intersection of B and C is identical to A; A, then, is the *meet* of B and C. (The operations of join and meet are partial.)

If $A_1, ..., A_n$ are parts of B_1, if $A_1 U...U A_n$ exists, and if $A_1 U...U A_n$ = B, then $\{A_1,..., A_n\}$ is a *finite dissection* of B.

Finally, a *binary structure* $\bar{W}$ = <W,π> is a *structure of bodies*, if, and only if, the following axioms obtain for all A, B, C and D in W:

1. A π A (Reflexivity).
2. I A π B and B π A, than A = B (Anti-symmetry).
3. If A π B and B π C, then A π C (Transitivity).
   (Note, then, that the part-relation gives a partial ordering of the bodies in W.)
4. If A and B have a common part, then they have a greatest common part.
5. If A and B have an envelope, then they have a least envelope.
6. If A is part of B and A ≠ B, then there is a body C in W such that B is the least envelope of {A, C}.
7. Every body has a least part.
8. Every body has a finite dissection of least parts.
   (Note, too, that axioms 7 and 8 commit the theory of bodies to atomism; by 7 every body contains at least one atom, and by 8 every body is composed of finitely many atoms.)

The Noll-Suppes account of part is more adequate for the present purposes than the standard idea of set-membership. But it is still somewhat too heavy-handed. We shall mention just two examples. (1) The axioms for the part-relation provide that properties true of all parts *always* compose. But what makes composition so interesting is that it is *sometimes* a fallacy, the fallacy precisely of supposing that what is true of all parts is true of the whole! The Noll-Suppes notion of part entails that composition is not a fallacy; and that, indeed, is too awkward a consequence. (2) Moreover, in the atomism of this theory there are additional difficulties. Intuitively, if a chain is pure gold so are its parts. But it is not required of *Suppesian* parts that the smallest parts of a chain be its links. In particular, the Noll-Suppes theory allows the *chemical* atoms of the links to be parts of the chain. As a consequence, a startling number of attributes which we could confidently expect to compose and divide fail to do so; and fallacies are ascribed where none there are.

3. *Are Wholes and Parts the Same as Burgean Aggregates and their Components?*

The notion of an aggregate and its components has been recently developed by Tyler Burge [28]. For current purposes, aggregates can be linked to first-order sets, i.e. sets containing only individuals as members. However, aggregates are importantly different from sets. For example, the empty set is *not* an aggregate and no singleton is an aggregate; from the point of view of aggregate-theory, a singleton of which the sole member is an individual just is that individual. As will shortly become clear, aggregates also exhibit sharp differences from mereological classes and Suppesian bodies. We turn now to the formal development.

In place of the set-theoretic notion of membership, given by the ε relation aggregate theory speaks of *componentiation*, i.e. being a component-member, symbolized by 'α'. Wherever x bears to y the α-relation, we could say, somewhat barbarically, that *x componentiates y*. The predicate expression, '**a**x,' for 'a is an aggregate,' can be defined thus:

(i)   **a**x ↔ ($\exists$z) ($\exists$y) (z α x. y α x. z $\neq$ y)

That is, x is an aggregate if, and only if, at least two objects are component members of it. A definition is also given for aggregate-abstraction:

(ii) $\hat{x}/\varphi x$ = df ($\imath$y) (x) (x α y ↔ ($\exists$z) (φz. x α z))

The counterpart of set-comprehension is;

(iii)   (y φ $\hat{x}/\varphi x$) ↔ ($\exists$z) (φz.y α z)

An explicit denial of the empty aggregate can be got from:

(iv)   ¬($\exists$ y) (y = $\hat{x}/x \neq x$)

Analogous to the principle of extensionality for sets we have:

(v)   (**a**z.**a**y) → (y = z ↔ (x) (x α y ↔ x α z))

where the antecedent reflects the intention of the theory that not everything be an aggregate. In fact, our next principle makes the point more precisely: only individuals are components or aggregates.

(vi)   x α y → Ix,

where individuals may be taken to be their own components:

(vii)   Ix → x α x. (z) (z α x → z = x)

The notion of an *individual* is defined by

(viii)   I(x) ↔ If ($\exists$w) (x α w)

It is clear at once that the α-relation is also quite different from the π-relation of the Noll-Suppes theory of bodies. Where, A π A holds, **a** α **a** does not (for A a body and **a** an aggregate). What is more, not all parts of an aggregate are components of it. The transitivity principle, if x α y and y α z, then x α z, does not obtain where y and z are aggregates. However, transitivity is provable in case x and y are just the same individual. In fact, given the joint assumption of x α y and y α z, it *follows* that x = y, and therefrom that x α z. So aggregates are not mereological classes or Suppesian bodies. Aggregates are entities that suggest themselves for the analysis of idioms in which there are plural constructions and mass terms, e.g. for such expressions as "the stars that presently make up the Pleiades galactic construction". Unlike sets, aggregates are physical entities in space-time, capable of action and of change, and susceptible of coming into and going out of existence. Unlike mereological classes and Suppesian bodies, not all parts of an aggregate are parts that *make up* the aggregate; no aggregate is its own member-component and no aggregate may be the component member of any aggregate (i.e. aggregates are always aggregates of individuals.) By virtue of the first of these three differences, one might be able to think of an iron chain as an aggregate, the ironness of which divides over its parts (i.e. its member-components, i.e. its links), and not worry about, e.g. the atomic parts of these components. They are parts that do not matter. And by virtue of the third of these three differences, aggregate theory avoids Russell-Zermelo problems of aggregates of all aggregates that are not component-members of themselves. In Burge's theory, predicates of aggregates are kept reasonably well-behaved and projectible. Aggregate-theory is rather well-suited, therefore, to our interest in composition and division. How, then, should the part-whole relation be taken for the purposes of a reasonable analytic understanding of the fallacies of composition and division? Not, certainly, as set theoretic membership or set-theoretic inclusion. And not as Suppesian body-parthood. The better prospect, though we do not claim perfection for it, is Burgean componentiation. In this fashion, a formal theory - the formal theory of aggregates - seems next-to-indispensable for a decent account of the traditional fallacies of composition and division.[16]

Chapter 18
The Fallacy of Many Questions

The objective of this paper is the analysis of what is traditionally called the informal fallacy of *many questions*, most commonly illustrated by the question (1) 'Have you stopped beating your spouse?'[1], which as Hamblin [80, p. 38] notes, seems designed to force ordinary non-spouse-beaters into admission of guilt. The traditional name may, as Åqvist [6, p. 75] suggests, be a misnomer.[2] Hamblin suggests that "Complex Question" seems more comprehensible. Though we will see that both names are misleading, for the time being we will continue to use the traditional one to represent our goal of analysis.

We wish to work towards considerations that would help in teaching students of logic - not only mathematical logic, but logic in the broader sense of evaluation of arguments - to be able to be in an initial position to (a) recognize the fallacy, or at least be aware of when it is likely to occur, (b) try to avoid the fallacy, or at least learn how to render it harmless, and (c) to begin to understand, at an adequate level of clarity, what is essentially fallacious about it, i.e. to understand how it could be a form of incorrect argument, and (d) to catch a glimpse of how the fallacy is effective as a dialectical manoeuvre.[3] But we stress - and here the reader must be careful to appreciate the precise, somewhat novel, character of the project - that the exercise is not one exclusively of pure formal logic, but in the application and adaption of formalisms to the pragmatic study of an informal fallacy.

Of course we might be inclined to reject the stipulation that a fallacy should always be thought of as representing a form of statement-based argument that is invalid or at least incorrect. Is it not a truism that informal fallacies are in some sense "informal"? It is argued in Woods and Walton [221] and Woods and Walton [242] that pragmatic factors are to be taken account of in the analysis of informal fallacies, and that the formalisms that can best be brought to bear may be non-standard. But the view that no formalism is at all applicable must be rejected. The study of the fallacies should not become so "informal" that classification of arguments into clear categories of 'correct' and 'incorrect' is waived.[4] Informal fallacies are more than merely propagandistic devices, rhetorical ploys, or psychological belief-modifiers.[5] They are first and foremost bad arguments. And bad in a way that must be studied in an adequate basis of theory as well as practice, so that an allegation of 'Fallacy!' can be rationally and objectively adjudicated, disputed, prosecuted, or defended.

1. SEMANTIC PRELIMINARIES

As we will see, Åqvist characterizes the fallacy as a question that has a *false* presupposition. But this move makes the fallacy material rather than formal. Just as formal logic does not tell us in general whether premisses are true or false, but only whether arguments are valid or invalid, even fallacy theory should not be construed so broadly that it should be expected to tell us generally whether statements are false or not. Moreover, the characteristic feature of 'many questions' is not that the presupposition of the question is simply false, but rather that it is somehow unwarranted or unwelcome in the context of the argument or disputation (even if it might turn out to be true). But these notions seem to take us even perilously further outside the scope of classical formal logic. Hamblin [80, p. 217] phrases this general limitation quite pointedly, in commenting on the thesis of Åqvist [6, pp. 74-5] that the fallacy is committed only by questions that have *false* presuppositions. One problem then is to see how the notion of the *unwelcomeness of a presupposition in a context of inquiry* can be formulated within a framework something like that allowed by the approaches of Åqvist, Hintikka, and Belnap and Steel to erotetic logic.

In what does the essential fallaciousness of 'Have you stopped beating your spouse?' consist? We propose that the question exemplifies an informal fallacy that is a significant error or pitfall of argument insofar as it attempts to speciously force the person to whom it is addressed into the position of signalling her acceptance of a proposition that is "unwelcome" to her, a proposition that would not be acceptable to her under the circumstances. The question 'Have you stopped beating your spouse?' in effect permits only two possible (simple and direct) answers, *yes* or *no*. But either answer implies what is unwelcome, we may presume, to the answerer, namely that she has beaten her spouse. A simple flow chart is suggestive.

The Fallacy of Many Questions 235

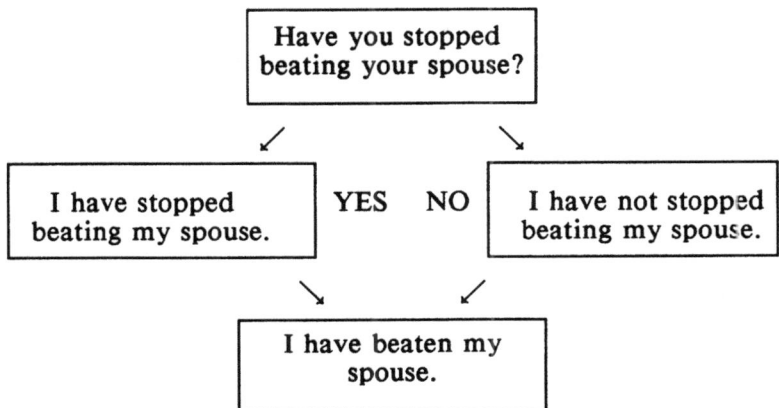

The answerer is trapped, because no matter which alternative she chooses, *yes* or *no*, she is committed to the same unwelcome implication. The fallacy is: what appears to be a genuine choice of alternatives is in reality a trap.

The above sketch represents what we will take to be the most plausible account of the gist of just what is fallacious about the alleged fallacy. The sketch does not explain the fallacy, it merely gives a plausible picture of the thrust of the intuitive fallaciousness of it. We now turn to a study of various details of the fallacy keeping this sketch in mind as a target, in order to try to bring out its deeper features more explicitly and exactly.

At the outset we must begin to recognize that there might be nothing fallacious about a question with a multiple presupposition, even though it may be reasonable to ask a questioner separate complex questions into smaller units. Moreover, there might be nothing wrong with a presupposition-containing question in relation to one specific context of disputation or inquiry, whereas the same question could be unreasonable relative to another context. For example "What did you use to remove your fingerprints from the gun?" could be an appropriate question following a confession of homicide, but an unreasonable question where no admission of connections with firearms or crimes has preceded.

2. THREE MAJOR METHODS

In this section, first we succinctly introduce a series of definitions of key formal concepts of erotetic logic of Åqvist [6], Belnap [15], and Hintikka [102], needed for the analysis to follow. The definitions are abstracted from important aspects of disputation and, as we will see, are formal stencils of complex phenomena of argument.

Åqvist's system is an epistemic-doxastic erotetic logic. That is, he reduces every expression of the form 'Is it the case that p?' to an expression of the form 'Let it be the case that either I know that p or I know that ¬p.' Hintikka's system is also an epistemic formal logic based on Åqvist's approach. However, we will omit the epistemic and imperative operators in what immediately follows - where this can be done without missing the point - reducing the illustrations to their underlying truth-functional structure.

A *whether-question* poses a number of alternatives, of which the answerer is supposed to select one. For example: Is she wearing the red dress or the green dress? Each of the alternatives is called a *direct answer*. Any statement implied by every direct answer is called a *presupposition* of the whether-question. Take the whether-question 'Is she wearing the red dress and the purple hat, or is she wearing the green dress and the puce shoes?' In effect, the question poses an alternation of two conjunctions, i.e., it says: $(R \wedge H) \vee (G \wedge S)$? thus the direct answers are '$R \wedge H$' and $G \wedge S$'. An example of a presupposition would be 'She is wearing a dress', because it is implied by both direct answers.

Belnap [15, p. 127] proposes that every question presupposes that at least one of its direct answers is true. Then he rules that the proposition that at least one of its direct answers is true is called *the* presupposition of the question. A question is called *safe* if its presupposition is logically necessary, *risky* if it is not safe. For example, 'Is she wearing the red dress or not?' is safe because its presupposition as '$R \vee \neg R$' is logically necessary. *Yes-no questions* are always safe because their presupposition consists in a pair of contradictory alternatives, e.g., the presupposition of 'Is snow white?' is 'Snow is white or snow is not white.'

The fact that the foregoing definitions are based on classical logic makes their applicability somewhat questionable in the present pragmatic context. For example, the following statement is a presupposition of 'Is she wearing the red dress or the purple hat?': she is wearing the purple hat or Plato was not an empiricist. For another example, suppose that the following is a direct answer to a certain question: if all men are mortal and Socrates is a man, then Socrates is mortal. Then the following statement is a presupposition of that question: if all men are mortal, Socrates is mortal, or if Socrates is a man then Socrates is mortal. These consequences are however no more paradoxical than the classical logic which yields the meaning of 'implies' in the above definitions. As Hintikka [102, p. 27] notes, the term presupposition may have to be treated, in the context of classical formal erotetic logic, as a mere *terminus technicus*. More then about pragmatics of erotetic conditionals as we proceed.

Let us see how the fallacy of many questions can be studied on the classical basis in (1): Have you stopped beating your spouse? Following Åqvist, let W = You have a spouse and have beaten him, and S = You have stopped beating him. Then the question says: (W ∧ S) ∨ (W ∧ ¬S)? But this is truth-functionally equivalent to the ordinary statement W. So (1) is risky. If the presupposition W is in fact false, it is impossible to give a true direct answer to (1) since W appears on both sides of the disjunction, (W ∧ S) ∨ (W ∧ ¬S). Thus the only sensible answer is to "correct" the question, perhaps by pointing out the falsity of W. So the fallacy arises where a question that is actually risky and moreover has a false presupposition is put in the guise of a safe "yes-no" question, according to Belnap [15, p. 128] and Åqvist [6, p. 66]. Syntactically, the question is safe, but semantically it is risky - a contradiction.

Hintikka [102, p. 28] treats (1) somewhat differently from Åqvist or Belnap. He notes that (1) has a presupposition the statement, '(∃x) (you stopped beating your spouse at x)' which in turn implies that before x you were beating your spouse. He describes such a question therefore as "notoriously loaded". It is useful to remember however that not all questions have that substantive presupposition are fallacious. 'Is she wearing the red dress or the purple hat?' need not be fallacious if its presupposition is not concealed. So Hintikka's account does not by itself explain what is precisely fallacious about (1). Moreover, not all risky questions are fallacious that (a) have a *false* presupposition, and (b) are in the form of a "yes-no" question (or perhaps other form of question that appears safe). 'Are you the student who sat at the back and asked a question yesterday?' may be unfallacious, say if the "student" was really a disguised teaching evaluator. Certainly this question is not fallacious in the same way that our initial sketch of (1) suggested an unfair manoeuvre of overly aggressive and deceptive questioning. It seems therefore that neither the Hintikka or the Belnap-Åqvist explanations are entirely satisfactory. So the Belnap-Åqvist account does not by itself explain the fallaciousness of (1) either.

Three major methods of transforming a risky question into a safe question are reviewed by Åqvist [6] and applied to the spouse-beating instance of (1). Let us look at these with a view to studying how far formal erotetic constructions can help us to further approach the fallacy. In the sequel we will reformulate (1) as (2), namely, 'Have you stopped beating your (one) spouse?', for as Peter Geach pointed out, logic does not exclude the possibility of a polygamous situation. (1) has more presuppositions than anyone has noticed!

A. HARRAH'S METHOD: REPLACE (1) WITH A PROPER YES-NO QUESTION.

This suggestion is to replace (2) with (3), 'Is it the case that you have one spouse and have beaten him and have stopped?' Letting S be as before, but now changing W to 'You have exactly one spouse and have beaten him', (3) asks: $(W \wedge S) \vee \neg(W \vee S)$? The second disjunct corresponds to the direct answer 'No', i.e., 'It is not the case that I have one spouse and have beaten him and have stopped.' We can see how this works by noting that (2) is a *fallacy* because instead of the *contradictory* alternatives posed by the proper yes-no question (3), it proposes merely *contrary* alternatives '$W \wedge S$' and '$W \wedge \neg S$' and is therefore risky, even though its resemblance to (2) may make it appear safe. Harrah's method simply corrects this deficiency in the obvious way.

One problem with Harrah's method however is that it does not allow the alternative '$W \wedge \neg S$' of the original question (2). That is, it does not allow the direct answer 'I have a spouse and have beaten him and have not stopped.' Thus it might seem more desirable to have a method that leaves all the original alternatives open.

B. THE WHETHER-WHETHER METHOD

A second method, one that overcomes the problem with Harrah's method, is simply to add the third alternative $\neg W$ to (2), or in other words transforming (2) into $\neg W \vee (W \wedge S) \vee (W \wedge \neg S)$? This question is safe, since its presupposition is a truth-functional tautology. Moreover, it has the advantage over Harrah's method that it preserves all the original alternatives of (2). The underlying principle in this transformation is what Åqvist calls the *Whether-Whether Method*: given a risky question composed of a set of alternatives $P_1 \vee P_2 \vee ... \vee P_n$, simply disjoin the negation of it, $\neg(P_1 \vee P_2 \vee ... \vee P_n)$, to itself, yielding $\neg(P_1 \vee P_2 \vee ... \vee P_n) \vee (P_1 \vee P_2 \vee ... \vee P_n)$. The result will always have the form of a truth-functional tautology and will thus be safe. We can see that this general principle is the method we have used here in transforming (2) by noting that $\neg W$ is the truth-functional equivalent of $\neg((W \vee S) \vee (W \vee \neg S))$. There is, however, one problem with the Whether-Whether transformation of (2). As noted by Prior and Prior [152], only $W \wedge S$ (*Yes*) and $W \wedge \neg S$ (*No*) are direct answers to (1), whereas the reply ¬W (I have no spouse whom I have beaten) is not so much an "answer" as a restructuring of or perhaps "avoiding" the question. As Åqvist [6, p. 70] puts it ¬W is only "allowed for" rather than being "called for" as a direct answer. Thus it would be nice to have a method to reflect this difference.

C. THE METHODS OF CONDITIONAL GUARDING

Another method of dealing with (2) is to break it down into two questions; (4) Do you have one spouse you have beaten?, and (5) If so, have you stopped? In the previous symbolism, (4) is just 'W?'. Therefore (5) might seem to have the form 'W ⊃ S'?. This is not quite right however. The question, more accurately, seems to attach only to the consequent and not strictly speaking to the whole conditional. The form might be better rendered as W ⊃ ?S, putting the question operator *before* the proposition it is meant to apply to. Another contrasting pair might make the distinction a little more apparent: (6) Is it true that if silver is metal, it conducts electricity? (7) If silver is metal, is it true that it conducts electricity? In the second case 'Silver is a metal' is a condition within the asking of the questions, and is not itself being questioned as part of a larger question. Note that 'Silver is not a metal' functions differently as an "answer" in each case. It does not affect the answer to (6) at all, one way or the other. However, if true, it tends to vitiate (7) altogether, ruling out either answer. Åqvist translates (2) into his symbolism as a conditional question as

(8) !(W ⊃ (K(W ∧ S) ∨ K(W ∧ ¬S)))

or equivalently

(9) !(¬W ∨ (K(W ∧ S) ∨ K(W ∧ ¬S)))

This can be contrasted with the Whether-Whether reconstruction of (1):

(10) !(K¬W ∨ K(W ∧ S) ∨ K(W ∧ ¬S)))

The difference here is reflected by the fact that W as antecedent of the conditional is not prefaced by a K-operator. Then Åqvist gives two methods for transforming (2) into a safe question. First, by the *Whately-Prior*[6] method, we transform (2) into

(11) W? ∧ (8)

Second, by the *Whether-if* method, we transform (2) into (8) by itself. According to Åqvist, this is a *conditional* method because he defines a conditional questions as follows:

$$?(p_1,p_2,...,p_i/q_1,q_2,...,q_j) =_{df} !((p_1 \wedge p_2 \wedge ... \wedge p_i) \supset (Kq_1 \vee Kq_2 \vee ... \vee Kq_j))$$

But we think the same point may be made more simply by noting that the scope of the ? operator in the *Whether-if* method is restricted to the consequent of the conditional. Then the *Whately-Prior* method amounts to the transformation of (2) into the two questions W? and W $\supset$?S.

The same kind of distinction we made between (6) and (7) is therefore the crux of what Åqvist wishes to show by this more refined method of imperative-epistemic transformations. He cites the examples below in the light of the discussion by the Priors.

(12) If you have a spouse you have beaten, have you stopped?

(13) If a bull were chasing you, would you climb a tree?

As the Priors put it: To say 'No bull is chasing me' would not excuse me from answering (13), whereas to say 'I have no spouse whom I have beaten' does excuse me from answering (2) and also puts an end to the inquiry.

Belnap and Steel [16] introduce the *conditional interrogative* (P/I) with *condition* P and *conditioned interrogative*, I. They define these notions in such a way [16, p. 103] that (P/I) calls for an answer only if P is true in a given interpretation of the interrogative, and I calls for an answer in that interpretation. For example, 'If you are going, are you taking an umbrella?' calls for an answer just in the case the respondent is going. Otherwise, as they put it, the question is inoperative.

Thus we can summarize the gist of the advice given by formal erotetic logic on how to deal with the spouse-beating question; rephrase it as a relativized conditional question, then if the condition 'You have a spouse you have beaten' is not met, the question can be treated as one that does not call for an answer.

This advice does not adequately explain what is fallacious about (2) however. Initially confronting the fallacy is not enough, nor have we seen clearly enough yet precisely how to identify what it is we confront that marks (2) off from more harmless questions. The sum total of this advice is that the question must be relativized to a certain condition. But we are nowhere told (i) how to identify the specific condition at issue and tell whether it is warranted, and (ii) precisely what form of conditional is to be the correct formal account of such relativization. The lesson appears to be that formal logic of the classical variety can only take us so far in the analysis of the many-questions fallacy.

3. A PRAGMATIC APPROACH

As we have seen, the current accounts of 'presupposition' in erotetic logic are essentially based on classical implication, and this basis can result in some unintuitive consequences. Belnap and Steel [16] suggest however that there are other possibilities. For example, they remark [16, p. 110] that the question, 'Is the present King of France bald?' could be parsed in a gappy way after Strawson, as well as in the classical way that they favour in their own account. In a free logic, we could say that 'The present King of France is bald' has a truth value only if there is now a unique French king - otherwise it is said to lack a truth value.

Still another account of 'presupposition' is epistemically oriented. Asking the question, 'How many bones are in a lion?' presupposes in this sense that the questioner does not know the answer but thinks that the respondent either knows or can find out. According to Belnap and Steel this last account of presupposition is *pragmatic* - it describes the questioner, the respondent, and the empirical context in which the question is asked. Therefore this account is outside the scope of the approach to erotetic logic pursued by all the erotetic studies we have so far looked at, including that of Belnap and Steel.

However, given the objectives of the present essay, a pragmatic approach is not only appropriate but positively required. For the simple fact that the presupposition of (2) is false is not by itself sufficient grounds for determining (2) as in any appropriate way fallacious. Rather the presupposition of (2) could be described as making the asking of (2) fallacious if that presupposition is not warranted or known to be true relative to the context of the interchange between the questioner and the respondent. Let us say that a presupposition of a question is *unwelcome* if, and only if, that presupposition is not established relative to a given point of a context of inquiry. We mean by 'established', known to be true by the participants in the inquiry.

When is a presupposition of a question unwelcome? The answer to this question takes us well out of the alethic framework of standard (classical) logic and puts us into the dialectical framework of Hamblin [80] where we need to think of an argument as a two-person or many-person game, composed of a set of participants, and a set of rules for making moves in the sense of advancing or retracting commitments in the sequential fashion characteristic of games [80], [81]. Thus there may be many different doctrines of the unwelcomeness of presuppositions depending on what sort of game one has in mind and also on the purpose and interpretation of the game, the form of argument the game is designed to model. Hamblin develops several of these formal games of dialectic that might be adapted to model varieties of unwelcomeness of presuppositions relative to a given game of argument. To me it seems clear that these games are the best basis for future research concerning unwelcomeness of presuppositions.

4. EPISTEMIC SETTINGS

For several of the fallacies, notably *petitio principii* and *ad ignorantiam*, an epistemic setting seems to bring out most forcefully the nature of the particular error of inference that is involved. And of course Hintikka [102] has developed epistemic logic as a powerful tool for the semantic analysis of questions. Although Hamblin prefers the dialectical to the epistemic approach because of worries about rationality assumptions, clearly Hintikka's new game-theoretic approach to epistemic logic has cleared the way for a synthesis of the two approaches. Perhaps the findings of section 1 could be enriched by looking to epistemic formalisms. A hopeful candidate is the Kripke [117] semantics.

The Kripke semantics models the idea that different propositions can be established (verified) at different points in a tree-like (branching) context of inquiry. The model is *cumulative*, in the sense that if a proposition is verified at a particular point, then it must remain verified at all subsequent points of the inquiry. In Woods and Walton [226], it is indicated how Kripke models can represent a useful pragmatic context for studying the fallacy of begging the question as a dialectical fallacy, and it is clear that it is also a very suggestive model to study the notion of unwelcomeness in a complex question.

We presume that given a question, we can put it into relation to a given Kripke model in such a fashion that we can think of the question being posed at some specific point (evidential situation) in the model. Each question presupposes a context of inquiry. Relative to that context, statements may be thought to be established (not unwelcome) or not established (unwelcome). The Kripke semantics, based on intuitionistic logic, models the idea that a given proposition can be established in a cumulative model of rational inquiry. The model is cumulative in the sense that, once a proposition is verified, it is always verified ever after in that model, i.e. there can be no retractions.

We remember that in classical logic, according to the Belnap-Åqvist explanation, the fallacy of (2) resides in the contradiction between (i) the supposition that (2) is risky, i.e. that the presupposition is contingent, and (ii) the supposition that (2) is safe. However, in the Kripke model, the presupposition 'Either you have a spouse whom you have stopped beating or you do not have a spouse whom you have stopped beating' cannot be regarded as always verified. That is, in the Kripke model, $A \vee \neg A$ is not a tautology, because intuitionistic negation, $\neg A$, does not mean that A is not verified at a given point but that A can never be verified at any point relative to a given point. A can be not-yet-verified but not yet ruled out either. By these lights, (2) is not safe at all.

What the Kripke semantics brings out about (2) is the point emphasized by Hintikka, namely that (2) is loaded in the sense that it has an unwarranted presupposition. What "unwarranted" means in relation to the Kripke model is very clear - a statement A may be said to be *unwarranted* or *unwelcome* just in case A is not verified relative to a given point in the model. By these lights, what is wrong with (2) is that the proposition that the answerer has a spouse that she has beaten is presumably not established relative to the context of the inquiry at the point at which the question (2) has been asked.

An extremely important property of the Kripke model is that is has the property, as we call it, of being essentially cumulative: if a proposition is verified at a given point than that proposition must remain verified at every accessible point. But surely the fallacy of many questions should be studied in non-cumulative epistemic contexts of inquiry as well. Suffice it to say that both cumulative and non-cumulative systems are constructed by Hamblin [80] in various attempts to model dialectical reasoning through formal dialogues, or games of logic. For our various purposes in studying the pragmatics of questions, it will be interesting to drop or retain cumulativeness conditions.

As is shown in Woods and Walton [226], the severest problems of Hamblin games as models of the fallacies concern difficulties of the deductive closure of commitments. For example, Hamblin considers a rule of the following form as a device to aid in the managing of commitments in the asking of questions: 'Question A, B, ..., X?' may occur only when $A \vee B \vee ... \vee X$ is already a commitment of both hearer and speaker. This rule would ban the asking of a loaded question where *the* presupposition in the Belnap-Åqvist sense, $A \vee B \vee ... \vee X$, is unwelcome to the hearer. So it would seem to deal with (2). But is does not entirely deal with objectionable questions like (2) because it does not deal with each individual presupposition of (2). For example, (2) has the presupposition $(W \wedge S) \vee (W \wedge \neg S)$: either you have a spouse you have beaten and stopped, or you have a spouse you have beaten and not stopped. But if commitments are not closed under implication, then even though the presupposition is equivalent to W, a participant in the Hamblin game could be committed to $(W \land S) \lor (W \land \neg S)$ but not to W or vice versa. So, for example, just because that participant is not committed to W, it doesn't follow that we can't ask her (2).

We already noted above that if we define a presupposition in the Belnap-Åqvist way, then any unsafe question will have an infinite number of presuppositions, because of classical rules like $p \rightarrow (p \lor q \lor r \lor ...)$. Now conjoin this to Hamblin's rule that a question should not be asked unless it is a commitment of both hearer and speaker, and it is required that the speaker and hearer each have an infinite number of commitments.

This problem is quite a general one for the study of the fallacies. What is needed is a conditional that allows for deductive closure of only those commitments that are closely related to a commitment at issue. The problem is that a cumulative system like the Kripke model has total closure of commitments, whereas the Hamblin games do not require any closure of even the direct consequences of commitments. The solution adopted in Woods and Walton [242] is to introduce a relatedness logic which allows that only propositions directly related to and implied by a commitment admit of closure. However, the problem is quite a general one for the study of conditionals, and we cannot deal with it here.

It seems that there is more than one aspect of (2) that could be described as fallacious, but so far none of these aspects by itself appears to exhaust the fallaciousness of (2). It is time to resolutely attempt a pragmatic overview of what could be wrong with (2) and similar fallacious questions.

5. LOADED AND MULTIPLE QUESTIONS

A question with an unwelcome presupposition is to be here called a *loaded* question. We shall mean 'unwelcome' in the sense that the presupposition may not be regarded as fairly established relative to a given dialectical game. Of course the Kripke model is one clear instance of such game. We define a *safe* question as one where the presupposition is a tautology according to the procedures for determining tautologies adopted by the participants of a dialectical game. We may leave it as an open question, relative to a given game, whether or not all safe questions are not loaded. Combining these two concepts, we get a partial but strong explanation of the fallaciousness of (2). The safe-appearing (2) is not only risky, but loaded.

The Fallacy of Many Questions 245

In addition to this trickiness however, it contains yet another snare for the unwary. It *forces* the intended victim to accept the unwelcome presupposition no matter which way he answers *yes* or *no*. Like the well-known frustrating questions of objective tests, it requires but does not contain an alternative 'None of the above'. There is more to the fallacy than its being really loaded when demurely offering the appearance of safety. Not only is it loaded, but *all* the chambers are loaded.

A safe question may be described as one that has alternatives that are logically exhaustive of all the possibilities. To answer it you must choose one.

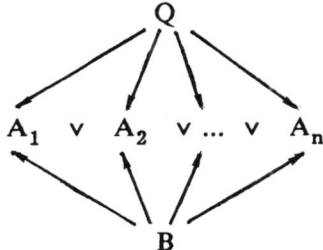

But no matter which one you choose, you may also be forced to choose some unwelcome proposition B, individually implied by each of the A_i. The deeper explanation of the essential fallaciousness is that Q appears safe because the A_i are logically exhaustive and consequently $A_1 \lor A_2 \lor ... \lor A_n$ is a tautology. Up to this point, Q is indeed "safe". But the deeper level of analysis represented by the third stage of the diagram reveals that the A_i themselves collectively contain a presupposition that is not a tautology. Each and every one of the A_i implies B. And B, as might happen, may be not only *not* a tautology, it may be unwelcome.

So it is a question of peeling off two levels of analysis. At the first level the presupposition is safe, but at the second level it is loaded. And the twist is that we can't remain at the first level, for B is a deductive consequence of *every* A_i at the first level. Thus there is a third factor. Not only first, does the question appear to be safe while in reality it is risky, but second, it is more than risky, it is loaded. But third, the fallacy is *coercive* in that each disjunct of its presupposition individually implies the proposition that is unwelcome to the answerer.

We now have a fuller account of the fallacy. It explains a good deal of what is really fallacious about the spouse-beating question. However, *this* fallacy, while plainly an egregious and therefore interesting one, is not the only fallacy that might be called, or that has been called "many questions" ("complex question", etc.).

In concluding his discussion, Åqvist makes an interesting point about the label 'Fallacy of Many Questions'; "... the label does not appear particularly appropriate in view of the fact that it misleadingly suggests that what is wrong about questions involving false presuppositions consists in their involving two or more independent questions" [6, p. 75]. Consequently, using the label 'Fallacy of Many Questions' might lead us to overlook the distinction between fallacious and merely multiple questions. Belnap and Steel make a kindred point in remarking that the fallacy of many questions is badly named. Hence something like 'Fallacy of False-Presupposition Questions' might be more to the point, in Åqvist's or Belnap's terms. In the context of the foregoing - at least as I think - more adequate analysis, perhaps it might graphically be called 'The Fallacy of Force-Loaded Questions'. Certainly, "many questions" can be ruinously misleading.

Indeed, the philosopher himself wrote at *De Sophisticis Elenchis* 167b39: "[fallacies] that depend upon the making of two questions into one occur whenever the plurality is undetected and a single answer is returned as if to a single question." This difficulty is a different sort of problem than the fallacy we have been attempting to analyse in (2), yet clearly it is one ingredient that helps to explain an important aspect of how the fallaciousness of (2) works. The way to deal with this particular difficulty is simply to separate the questions, as Aristotle himself observes at *De Sophisticis Elenchis* 181b1: "To meet those refutations which make several questions into one, one should draw a distinction between them straight away at the start. For a question must be single to which there is a single answer, so that one must not affirm or deny several things of one thing, nor one thing of many but one of one." We should note emphatically however the risk of terminological confusion - or better, the fact of it - in that the historical progenitor of "many question" in the logic texts, namely this account from *De Sophisticis Elenchis*, is clearly not the whole story of what is fallacious about (2). What we have here is a phenomenon familiar in the fallacy domain, the evolution of the name to cover something quite different evidently from Aristotle's original account. We can see why the name "many questions" was originally appropriate, and has now, over centuries of pedagogical employment, become curiously inappropriate, at least in large part. Should we follow tradition and, at risk of adopting a possibly misleading label, continue to call our main analysis above an account of The Fallacy of Many Questions? Or should we boldly offer 'The Fallacy of Force-Loaded Questions?' Perhaps it doesn't

The Fallacy of Many Questions 247

matter. Our own inclination is to stick with tradition even if this commits us to some historical explanations of the apparent oddity of the traditional term. The important thing is to be clear on the distinction between Aristotle's multiple-questions "fallacy"[7] and the spouse-beating fallacy we have now opted to call Many Questions.

Another deficiency in question-asking related to the fallacy of (2) is suggested by an illustration of Aristotle's that does not fit any aspect of questions we have so far studied: "supposing A to be good and B evil, you will, if you give a single answer about both, be compelled to say that it is true to call these good, and that it is true to call them evil and likewise to call them neither good nor evil (for each of them has not each character), so that the same thing will be both good and evil and neither good nor evil." This too is an interesting form of argument from the point of view of erotetic fallacies but it appears to be a distinct fallacy from Many Questions. Let us call it the Black and White Fallacy. This is a form of whether-question, where the set of alternatives of the presupposition is incomplete, not in the sense of being not safe, but in the sense that it does not fairly represent all the reasonable alternatives. Thus to ask someone whether a zebra is black or white is to introduce a new dimension of unfairness of presupposition into the asking of questions. To ask 'Black or White?' is to omit some needful alternatives in the disjunctive presupposition B ∨ W. The alternatives should be: 'Black but not White,' 'White but not Black,' 'Both Black and White,' or 'Neither Black nor White.'

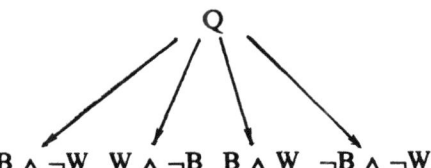

Now since (B ∧ ¬W) ∨ (W ∧ ¬B) ∨ (B ∧ W) ∨ (¬B ∧ ¬W) is a tautology, the question is not only safe but introduces sufficient discriminations amongst the alternatives to avoid the fallacy. How the Black and White Fallacy is to be analysed remains an open question, since the job requires some structure that could serve as a basis for a method of deciding when a set of alternatives of the presupposition of a question is adequately representative of the fineness of discrimination needed relative to the context.

One way of dealing with the Black and White Fallacy is by means of the method of relativized conditionalization of questions of Belnap and Steel. They propose that a question like 'If it was murder or suicide, which one was it?' can be parsed as a conditional question so that if it was neither suicide nor murder the question may be regarded as inoperative. We do think that this method provides an analysis of what is precisely fallacious about the Black and White Fallacy, but is does offer one way of coping with that fallacy.

However this fallacy is to be analyzed,[8] it is fitting to close a practical discussion of it by emphasizing its importance in actual argumentation by citing some items in a long an entertaining list of questionable questions compiled by Fischer [64, p. 10f.]. The following are actual titles of works published by reputable historians: *Napoleon III: Enlightened Statesman or Proto-Fascist?, The Abolitionists: Reformers or Fanatics?, Plato: Totalitarian or Democrat?, The Dred Scott Decision: Law or Politics?, Ancient Science: Metaphysical or Observational?, Feudalism: Cause or Cure of Anarchy?* As Fischer points out, these questions suffer from a wide variety of faults of shallowness and simple-mindedness. But whatever else their faults, they are splendid examples of the Black and White Fallacy. Finally, (2) has yet another aspect to its analysis not yet brought out. This has to do with negation in certain kinds of temporal expressions of continuity. It is clearly brought out by noting the analogy between (2) and the Megarian paradox of the horned man.

> What you have not lost you still have.
> You have not lost horns.
> Therefore, you still have horns.

W. Kneale and M. Kneale [114, p. 114] offer the following explanation. The second premiss involves a presupposition, i.e. 'You once had horns'. And that second premiss may be negated either in a restricted way that accepts the presupposition or in an unrestricted way without acceptance of the presupposition. The same sort of explanation could be applied to (2). 'You have not stopped beating your spouse' could be negated in a restricted way that accepts that you had once beaten and are continuing to beat your spouse. Or it could also be negated unrestrictedly in a way that does not accept that presumption, even if this second interpretation might be the less likely one to be usually taken up. Whatever the correct explanation of what is sophistical about the non-erotetic horned man paradox, its evident similarity to (2) brings out a new element of the Fallacy of Many Questions.

This aspect of (2) is especially remarkable in that it suggests that negation plays an important role in Many Questions and related fallacies, and this is an aspect that requires further study. The suggestion that different kinds of negation are involved again underlines the wisdom of a pluralistic approach. Neither the intuitionistic negation nor classical negation are therefore fully adequate to dealing with all relevant aspects of (2).

5. CONCLUDING REMARKS

We have discovered a number of distinct aspects of (2), each of which is of pragmatic relevance in understanding the fallaciousness of (2) as a whole, and how the parts of it work: (i) (2) is a question that appears safe but is risky, (ii) (2) is a loaded question; (iii) (2) is force-loaded - it leaves the answerer no choice; (iv) (2) is a multiple question; (v) (2) fails to make finely enough discriminated alternatives; (vi) (2) has an ambiguity in negation like the horned man fallacy. Moreover, we have argued that some of these six factors need not be in themselves fallacious. Importantly, none of these factors taken by itself - so we have argued - explains what is fallacious about (2). Clearly all of the six factors are elements of the fallaciousness of (2), but we leave it open here just precisely how these elements are to be weighted in a mapping of the overall geography of what we are calling the fallacy of many questions. Perhaps it is reasonable to suggest that the first three factors might be somewhat more significant as main elements of what is generally fallacious about (2) as an instance of the many questions fallacy.

It is also suggested by the foregoing analysis that we should reject any assumption that either asking loaded questions or multiple questions is in itself somehow intrinsically fallacious. This warning is encouraged by the observation that sometimes it is useful and reasonable to formulate and ask both kinds of questions.

Plainly, however, one needs to be careful in these regions, since the device of cleverly combining loaded and multiple questions can unquestionably be a powerful weapon in the sophist's arsenal. Such techniques, while not strictly speaking fallacious *per se*, can be trickily employed, and are devices that the student of fallacies should be familiar with.

An objection to our analysis is the following. If there is nothing fallacious *per se* about loaded or multiple questions, is it not the case that sometimes (2) is perfectly non-fallacious? If in fact the addressee of (2) is a known spouse-beater, surely the presupposition is a reasonable one. Hence surely, in that specific context, (2) is non-fallacious. The above analysis rules that the many-questions fallacy occurs when (i), (ii), (iii), (iv) and also perhaps the other two conditions are met. In this situation, the required conditions are met, but the question is not fallacious.

The reply is that according to the analysis we offer here, (2) is not always fallacious. We must remember that the analysis given here is always relative to the given context of inquiry or disputation. If all parties agree that the addressee is a spouse-beater, including that very person herself, then relative to that specific context of inquiry, (2) is not fallacious. Reason: condition (ii) is not met, because (2) is, in that specific context, not loaded.

In other words, according to the present analysis (2) can be fallacious or not, depending on whether or not the required conditions are met. This may represent something of a departure from tradition, according to which (2) is thought to be a fallacy. By our analysis, it is not always fallacious and can, in this one instance at least, be a reasonable question to ask.

Another objection is that in the everyday practice of real answering procedures (2) is perfectly non-fallacious because the obvious answer "I have never beaten my spouse", is perfectly correct. In ordinary reasoning, unlike formal erotetic framework, we allow the question to be asked. But then we answer it by questioning or rebutting that presupposition. After all, isn't that the natural response? One is not, in real argumentation, forced to tamely answer 'yes' or 'no' to (2).

This objection reminds us that the method of conditional guarding of II.C offers a way out of being forced to answer 'yes' or 'no'. If so, then how can it be that, according to the analysis above, the answer is left no choice? Hence the analysis must be wrong.

In reply it should be conceded that indeed the best strategy for the non-spouse-beater, or at least the addressee who does want to be acknowledged as a spouse-beater, is to rebut the presupposition W. The technical problem remains however - as we saw in section II - of precisely the best way to carry out this strategy.

The objection runs deeper than this however. It postulates that rebutting the presupposition W is a correct answer, or at least rightly should qualify as such in the everyday practice of argumentation. The objection rejects the formal erotetic approach of requiring direct answers to a yes-no question. After all, outside of formal dialectical games and other formalized structures, are we ever really *forced* to answer questions directly? How realistic are these erotetic models to the pragmatic question of whether (2) is a fallacy?

In reply, it should be pointed out that there are real-life situations where the addressee of a question is not offered the option of challenging the presupposition of a yes-no question, or a whether-question with a finite set of alternatives where the respondent is required to select one. The so-called "objective" examination questions, or multiple-choice questions, are sometimes alleged to be precisely of this nature by students required to answer them. It is as though the question was formulated in this way: You have stopped beating your spouse - answer 'yes' or 'no'. Similarly, in court occasionally a witness is required by the magistrate to directly answer the question 'yes' or 'no', and not to evade it or ask to have it reformulated. According to our analysis, this procedure need not be fallacious, although it may be if the presupposition is unwarranted relative to the inquiry.

So the point should be made that there are actual and not merely technically constructed situations in which the answerer is forced to give a direct answer to a question like (2). At least she is forced in the sense that no third alternative, or way of avoiding the question, is offered. In other ordinary situations however, it is clear that the best way to deal with the obviously outrageous question (2) is to rebut the presupposition W. It should be added however, that (2) is a study specimen, whereas more complex or subtle questions of the same form as (2) may actually occur in argumentation and be, for various reasons, harder to deal with.

Thus condition (iii) is a key element in the fallaciousness of (2). Asking someone "Are your still beating your spouse or not, if in fact you have ever done so at all?" is not fallacious, at least so we think, because it leaves the non-spouse-beater a clear way out. A third alternative is nicely offered. All you have to say is: "No I never did so at all." The deceitful strategy of (2) however - transparently obvious though it is in this particular amusing example - lies in the failure to offer the answerer this way out. That is not to say however, that in real life she won't take it anyway! (2) in many ordinary situations does not literally force the answerer to acknowledge her spouse-beating. But by not offering an alternative to options that have this presupposition, its structure is such as to unfairly restrict the options offered some answerers.

Chapter 19
Question-Begging and Cumulativeness in Dialectical Games

In two recent articles [234], [226], the authors argue that *petitio principii* is better understood as an epistemic rather than an alethic fallacy. In another article [227] we seek for an epistemic analysis of the *petitio* based on the von Wright-Geach notion of entailment.

However, other writers, e.g., Hamblin [80], have serious reservations about the usefulness of epistemic logic for the exploration of the fallacies and are themselves inclined to think that formal dialogues (games of dialectic) offer a more fruitful approach. Still, it has long seemed to us that dialectical theory can flounder in perplexities that are just as formidable as those of epistemic logic, and that the analysis of circular argumentation affords striking confirmation of the point. In this paper we further pursue solutions to some of these problems. It is our hope that we may move some steps beyond our efforts of 1978.

The dialectical perspective has strong ties historically with the development of doctrines of the fallacies, and these ties are documented by Hamblin [80] in enough detail to show how powerful the dialectical tradition has been, never mind its contemporary neglect. In western philosophy the dialectical tradition originates in the Platonic dialogues and in Aristotle's attempts in the *Topics* and *De Sophisticis Elenchis* to formulate rules of argumentation in a quasi-formal manner and to describe and propose solutions for certain "sophistical refutations". In the Middle Ages, from the thirteenth through the sixteenth century, we find numerous treatises by such influential logicians as Burley, Buridan, Strode, Albert of Saxony, Paul of Venice, all dealing with a dialectical game call Obligation. In Obligation, the task of the opponent is to draw out various impossibilities from a *positum* [position; argument] which is to be upheld by the respondent.

Green and Hamblin [78] point out that although Obligation is based on the debates of the Greeks, it is much more consciously a classroom exercise, and that the *positum* is often a quite ordinary false proposition - such as 'Socrates is black' - of no remarkable philosophical significance. It is not the content, but rather the

254   *Chapter 19*

form of the proposition in certain inferences that would seem to be the focus of dialectical interest. For example, if the *positum* is 'Socrates is black,' then the opponent might put forward the proposition [*propositum*] 'Socrates and Plato are the same colour' which the respondent must either concede or deny, according to the following rules: (1) consistency with the *positum* and previous replies must be maintained; but (2) if there is no question of inconsistency, the respondent must tell the truth. In the present example, the *propositum* is consistent with the *positum*, so the respondent must follow (2) and concede the *propositum*. If next the respondent is confronted with 'Plato is white', he must deny it, by (1), since it is inconsistent with the *positum* and the recently conceded *propositum*.[1]

Actually the dialectical study of argument has not been totally overlooked in recent times, as the following recent developments show.

1. THE HAMBLIN GAME (H)

Of special interest in studying the *petitio* is the 'Why-Because-System-with-Questions' of Hamblin [80, p. 265 ff.]. There are two participants, White and Black, each of whom has a commitment-store containing a finite set of statements. Each participant must add or delete commitments according to the rules given below. The language of this game (H) is basically first-order, restricted to a finite set of atomic statements. Axioms of the system are contained in the initial commitment-stores of both participants. White moves first, Black responds, and each continues in turn to make one move at a time. The capital letters S, T, U... are variables of the metalanguage for statements.

Locutions may consist of the following types:

(i) ⌜Statement S⌝ or, in certain special cases, ⌜Statements S, T⌝.
(ii) ⌜No commitments S, T, ..., X⌝, for any finite number of statements S, T, ... X (one or more).
(iii) ⌜Question S, T, ..., X?⌝, for any number of statements (one or more).
(iv) ⌜Why S?⌝, for any statement S other than a substitution-instance of an axiom.
(v) ⌜Resolve S⌝.

Two categories of rules are given; *locution rules* and *rules of commitment-store operation*.

Question-begging and Cumulativeness in Dialectical Games   255

Locution Rules:

S1. Each speaker contributes one locution at a time, except that a 'No commitment' locution may accompany a 'Why' locution.

S2. ⌜Question S, T, ..., X?⌝ must be followed by

    (a) ⌜Statement⌝ ¬(S∨T∨...∨X)⌝ ('¬' for negation)
or (b) ⌜No commitment S∨T∨...∨X⌝
or (c) ⌜Statement S⌝ or
       ⌜Statement T⌝ or
       ——————— or
       ⌜Statement X⌝

or (d) ⌜No commitment S, T, ..., X⌝

S3. ⌜Why S?⌝ must be followed by

    (a) ⌜Statement ¬S⌝
or (b) ⌜No commitment S⌝
or (c) ⌜Statement T⌝ where T is equivalent to S by primitive definition.
or (d) ⌜Statements T, T ⊃ S⌝ for any T.

S4. ⌜Statements S, T⌝ may not be used except as in 3(d).

S5. ⌜Resolve S⌝ must be followed by

    (a) ⌜No commitment S⌝
or (b) ⌜No commitment ¬S⌝.

Commitment-store operation:

C1. ⌜Statement S⌝ places S in the speaker's commitment store unless it is already there, and in the hearer's commitment store unless his next locution states ⌜¬S⌝ or indicates 'No commitment' to S (with or without other statements); or, if the hearer's next locution is ⌜Why S?⌝, placement of S in the hearer's store is suspended until the hearer explicitly or tacitly accepts the proferred reasons (see below).

C2. ⌜Statement S, T⌝ places both S and T in the speaker's and hearer's commitment stores under the same conditions as in C1.

C3. ⌜No commitment S, T, ..., X⌝ deletes from the speaker's commitment store any of S, T, ..., X that are in it and are not axioms.

C4. ⌜Question S, T, ..., X?⌝ places the statement Sv Tv ...vX in the speaker's store unless it is already there, and in the hearer's store unless he replies with ⌜Statement ¬(SvTv... vX)⌝ or ⌜No commitment SvTv ... vX⌝.

C5. ⌜Why S?⌝ places S in the hearer's store unless it is there already, or he replies ⌜Statement ¬S⌝ or ⌜No commitment S⌝.

It is of some importance that in (H) commitment-stores are not closed under the classical logical operations. Hamblin considers requiring that the statements in a commitment-store be *consistent* [80, p. 263f.] but rejects this because "consistency presupposes the ability to detect even very remote consequences of what is stored, and this would itself make nonsense of certain kinds of possible dialectical application". On the other hand, Hamblin suggests [80, p. 264] that certain "very immediate" consequences of a commitment might also be regarded as commitments in a given system. The general idea therefore is that (H) can be regarded as a base system upon which closure requirements of various strengths might be built up for various purposes of application.

Another important provision of (H) is that the retraction of commitments by a participant is allowed. That is, statements may be deleted from, as well as added to, commitment-stores at appropriate moves. Particularly difficult questions of degrees of closure of commitments under logical operations concern retraction. What is to happen if a participant retracts commitment to T or even replaces it by ¬T when he is committed to both S and ⌜S ⊃ T⌝ ? Again, Hamblin would have it that specific rules need to be laid down in specific systems to deal with this sort of situation.

Hamblin stresses that a commitment is not to be thought of as a "belief" of the participant who has it, and he disavows any implication that the interest or point of commitment-stores is psychological. It is also well to notice that Hamblin [80, p. 260ff.] develops a version of Obligation that does not allow for retraction of commitments.[2]

Hamblin [81] systematizes various critical options in quite a general way by constructing seven basic types of dialectical

Question-begging and Cumulativeness in Dialectical Games   257

system, all "information-oriented"; that is, their purpose is to exchange information between the participants. In Hamblin's System 2, for example, all logical consequences of commitments are also commitments. What this means in effect is that System 2 contains a rule that is a perfect analogue of the Hintikka "logical omniscience" rationality assumption. Not surprisingly, such a strong rule leads to problems for the formulation of realistic and workable mechanisms for the retraction of commitments. If a participant retracts T, but T is a consequence of S, should he be regarded in System 2 as having retracted S as well? The trouble is that is may not be possible to single out what these statements are that ought to be retracted. If S and S' together imply T but neither does it by itself, we have no obvious way in general of knowing which os S or S' should be retracted.

How do circles occur in (H)? The two most elementary forms of circular reasoning may be represented by the following dialogue-sequences. A, B, and C are atomic statements in the dialogues that follow.

	WHITE	BLACK
I	(1) Why A?	Statements A, A ⊃ A

	WHITE	BLACK
II	(1) Why A?	Statements B, B =$_{df}$A
	(2) Why B?	Statements A, A =$_{df}$B

Further sequences still, though not mentioned by Hamblin, also represent kinds of circular argument.

	WHITE	BLACK
III	(1) Why A? (2) Why B⊃ A	Statements B, B ⊃ A Statements A, A ⊃ (B ⊃ A).

	WHITE	BLACK
IV	(1) Why A? (2) Why B?	Statements B, B ⊃ A Statements A, A ⊃ B

The latter sequence represents a paradigm of circular argument, and Woods and Walton [226] call an exchange having this form a *circle game.* In a circle game it is possible to have a third step, C, intervening between the beginning and the end of the circle, as follows.

	WHITE	BLACK
V	(1) Why A?	Statements B, B ⊃ A
	(2) Why B?	Statements C, C ⊃ B
	(3) Why C?	Statements A, A ⊃ C

Then the same form of sequence can be carried through to n steps, allowing for many intervening steps before the circle is closed by Black.

	WHITE	BLACK
VI	(1) Why A?	Statements A_1, $A_1 \supset A$
	(2) Why A_1?	Statements A_2, $A_2 \supset A_1$
	. . .	
	. . .	
	(k) Why A_{n-1}?	Statements A_n, $A_n \supset A_{n-1}$
	(k+1) Why A_n?	Statements A, A ⊃ A_n

The second and third sequences that we looked at above may be likewise expanded. Though analogous remarks could be made about these other two kinds of game, in what follows we confine our observation to circle games.

Hamblin [80, p. 268f.] proposes two rules that block *petitio* in (H). The first is a rule for when the 'Why?' proposer is regarded as inviting his opponent to *convince* him:

(W)   ⌜Why S?⌝ may not be used unless S is a commitment of the hearer and not of the speaker.

otherwise 'Why?' is academic. The second rule is specifically designed to block circular reasoning:

(R1)   The answer to ⌜Why S⌝ ?, if it is not ⌜Statement ¬S⌝ or ⌜No commitment S⌝, must be by way of statements that are already commitments of both speaker and hearer.

How then do (W) and (R1) block *petitio* in (H)? To see how, consider the circle game, IV. When Black responds ⌜B, B ⊃ A⌝ at (1), it is required by (R1) that both statements A and B ⊃ A be commitments of both Black and White. Thus at (2), White's move ⌜Why B?⌝ is illegitimate by (W). So we can see that, given (W) plus (R1), circle games can never be played in (H).

Nonetheless, it is shown by Woods and Walton [80, p. 226] how sequences that may be circular can be constructed even in (H) + (W) + (R1).[3] These problematic sequences are not definitively demonstrable as circular arguments, but it is also not demonstrable that they are not circular; and indeed there is some conflict of intuitions on the matter. However, it is shown by Woods and Walton that if circularity is defined in a certain epistemic way in a Kripke model, circular dialogue-sequences can definitely be constructed in (H) + (W) + (R1): [226, p. 83]. It is perhaps too soon to tell exactly what has gone wrong with Hamblin's method for blocking the *petitio*, but a large part of the trouble appears to stem from the fact that (H) is not even "partially" closed under logical consequence, and that this makes for problems concerning the retraction of commitments in (H).

2. RESCHER DIALECTICS

Rescher [158] develops the thesis that dialectic, the study of rational disputation, can be instructive for epistemological controversies and that indeed it provides a certain sort of model of epistemic reasoning. The core of dialectic is formal disputation, which Rescher takes to be the study of a disputation having a set of rules and three participants, the proponent, opponent and determiner (referee or judge). In Rescher's version, there is an asymmetry in the roles of the first two participants, much like in Obligation. The proponent, who opens the disputation, is to maintain a thesis T by adducing supporting considerations or grounds in favour of T. The opponent attempts to rebut these moves by means of counter-assertions and challenges. The *burden of proof* lies with the proponent - his every move involves some categorical assertion. The opponent need make no positive assertions or claims. His sole function is that of challenging the proponent's claims.[4]

Three types of move are allowed to the proponent in a Rescher game: (i) *Categorical assertion*: !P, for "P is the case" or "P is maintained by the assertor", (ii) *Cautious assertion*: †P, for "P is compatible with everything you (the adversary) have said", (iii) *Provisoed assertion*: P/Q, for "P generally (usually, or ordinarily) obtains provided that Q". (Other readings of P/Q are "P obtains in all (most) ordinary circumstances (possible worlds) where Q does", or "Q constitutes *prima facie* evidence for P".) It is clear that this slash-relation is the key element in applying Rescher's dialectical framework to any study of argumentation.[5] The following characteristics of the slash are given: (1) P/Q is consistent with ¬P/Q&R; hence P/Q is not an implication relation. (2) It follows that *modus ponens* does not hold for the slash [158, p. 7]. (3) One of the key aims of disputation is the building up of an epistemically frail conclusion from relatively firm premisses, yet in a deductively valid argument, the conclusion cannot be weaker than the weakest premiss. So deductive implication is a stronger relation than the slash. (4) Moreover, according to Rescher [58, p. 7] the slash relation is not that of probability. (5) Transitivity fails: if P holds in most cases where Q does, and R holds in most cases where P does, it need not follow that R holds in most cases where Q does. (6) The law of contradiction is abandoned [58, p. 63]; the maintenance both of P, relative to a consistent basis Q, and of ¬P, relative to Q together with something else that is compatible with Q, is an open possibility. On the other hand both (P & ¬P)/Q and the combinations of P/Q with ¬P/Q are both blocked on strictly logical grounds. (7) The principle, *ex falso quodlibet*, that anything follows from an inconsistency does not hold in dialectical frameworks. [58, p. 64].

How then does the *petitio* arise in a Rescher disputation game? Two examples are given [58, p. 20] of sequences where the proponent reasserts something he has "effectively" asserted before.

(A) <u>Proponent   Opponent</u>
!P   †¬P
!P

(B) <u>Proponent   Opponent</u>
!P   † ¬P
!¬¬P = !P

The second would appear to be an *equivalence petitio*.[6] A third sequence given by Rescher [58, p. 20] would appear to constitute a *dependency petitio*.[7]

(C) <u>Proponent   Opponent</u>
!P   †¬P
P/Q & !Q   †¬Q
Q/P & !P

The problem with this approach is that though there is need for a "blockage rule" to stop these circles, Rescher nowhere precisely formulates a device adequate to block cases (B) and (C) above. In effect, therefore, Rescher's theory of formal disputation is confronted with the same difficulty as Hamblin's dialectical theory of argument. The problem of how to block circles is not solved.

We are led, then, to three fundamental questions. When is a circle fallacious? What does it mean to say that a circle occurs in *argument*? And what is wrong with arguing in a circle, when it *is* wrong? Neither Rescher nor Hamblin really addresses these questions with reference to their respective dialectical systems. But by attempting to formulate blockage rules for circles, they do lay some foundations for further investigations.

The slash is a new item in the logical repertoire. It is not deduction or induction of any recognizable sort (who would wish it to be?) nor is it a relatedness implication,[8] for the relatedness $\rightarrow$ is essentially a deductive relation. The slash is a binary relation on statements, and that is about all we know about its logic in a positive way. It would seem then that Rescher's approach would open the way for the possibility of a *pluralism* of conditional binary evidentiary or justificational relations that can crop up in dialectical argument. It seems to us unreasonable to wholly deny to Hamblin the use of deductive and inductive conditional relations in certain sorts of dialectical games, but no doubt Rescher is right in holding that in many realistic dialectical contexts, the appropriate evidentiary relations are weaker than, or at least different from, either deductive implication or its inductive counterparts. Various different kinds of dialectical games well may answer to the practical model of argumentation that one has in mind.

3. THE MACKENZIE GAME DC

Mackenzie [127] has constructed a variant game designed to study the *petitio*. Like (H), the game DC has similar rules for locutions, withdrawals, questions, challenges, and so forth. A proposition A is said to be *under challenge* for a participant if his opponent is committed to ⌜Why A?⌝. Mackenzie's innovation is that DC is not cumulative for statements, but is instead cumulative for challenges. The commitment rules provide no way of eliminating one's commitment to a challenge once incurred. Also, DC is cumulative with respect to immediate consequence conditionals because of the following two rules.

R_{Imcon}: A conditional whose consequent is an immediate consequence of its antecedent must not be withdrawn.

262 *Chapter 19*

$R_{Log\ Chall}$: A conditional whose consequent is an immediate consequence of its antecedent must not be challenged.

There is also a rule that is especially significant for the *petitio*; it is R_{Chall}.

R_{Chall}: After ⌜Why A?⌝, the next event must be either
(i) ⌜No commitment A⌝; or
(ii) The resolution demand of an immediate consequence conditional whose consequent is A and whose antecedent is a conjunction of statements to which the challenger is committed; or
(iii) A statement not under challenge with respect to its speaker (i.e., a statement to whose challenge its hearer is not committed).

The game DC allows for chains of challenges. If Bob says ⌜Why A_o?⌝ and Wilma replies ⌜A_1⌝, both are committed by the rules to ⌜A_1⌝ and If ⌜A_1 then A_o⌝. Bob could then challenge ⌜Why A_1?⌝ and Wilma might reply ⌜A_2⌝. And so forth.

The method of banning circles is similar to Hamblin's. Consider the sequence,

n Bob: Why A?
n + 1 Wilma: A

By the commitment rule for challenges, A is under challenge with respect to Wilma at n. Hence by clause (iii) of R_{Chall}, Wilma's move at n + 1 is illegal.

Now let us see how DC is supposed to succeed where (H) failed in handling the Woods-Walton dialogue-fragment, VII, that caused the trouble for Hamblin's proposed blockage rules of (W) + (R1). In DC, the Woods-Walton fragment is reconstructed by Mackenzie as follows. The commitment-store of Wilma is {B, If A then B, If B then A}. The commitment-store of Bob is {A, B, If A then B, If B then A}.

n Wilma: Why A?
n+1 Bob: B
n+2 Wilma: A
n+3 Bob: C
n+4 Wilma: Why B?
n+5 Bob: A
n+6 Wilma: B

Mackenzie judges that Bob begged the questions at n+5, and that the rules of DC reflect the error of his move - Bob broke R_{Chall}

by replying to a challenge with a statement that was under challenge with respect to him at that stage. A was under challenge with respect to Bob in virtue of Wilma's move at n.

Since DC is cumulative with respect to challenges, Bob has replied to a challenge with a statement that must *always* remain under challenge for him. He has transgressed R_{Chall}. But has he committed a *petitio*? Well, it might seem that he has in a fashion, but there is also the irksome fact that Wilma had *accepted* A at n+2. And given therefore that Wilma is committed to A from that point, it is not an unreasonable move for Bob to use A at n+5 as a response to Wilma's challenge ⌜Why B?⌝ at n+4. True, Wilma challenged A at n, but then later at n+2 she committed herself to A. And surely any statement that she has committed herself to is a reasonable basis from which her opponent should attempt to extend her commitments by argument. In short, a fair case can be made out for arguing that Bob's move at n+5 is not really a *petitio*, nor that is it a fallacious or illegitimate move, nor even that it should be illegal in a dialectical game designed to model the *petitio*.

The key reason that n+5 has the appearance of a *petitio* in DC is that challenges are *cumulative*. Hence Wilma's challenge. ⌜Why A?⌝ at n is inerasable. Once Wilma has made the challenge, Bob cannot ever use A as a basis for argument with her, even if she in fact commits herself to A at some subsequent point. Is this convention entirely fair or non-arbitrary for all contexts or argument? It is difficult to see why it should be so regarded. Thus Mackenzie's conclusion [127, p. 10], that cumulativeness is not the crucial factor in the *petitio*, seems to us somewhat evasive. What has happened in DC is that cumulativeness of *commitments* is replaced by cumulativeness of *challenges*, so the locus of the circle-arresting mechanism is shifted. But cumulativeness is not given up entirely, nor is the possibility of *petitio* in the Woods-Walton dialogue-segment resolved satisfactorily or decisively. Bob has broken a rule of DC at n+5 but has he committed a *petitio*? The best answer still seems to be: Yes or No. Yes, if challenges are cumulative and if *petitio* is the fallacy of replying at some stage to a challenge with a statement that is under challenge at that stage. No, if challenges can be retracted by subsequent commitments, or if there is a difference between the *petitio* and the use, as a basis for argument of some statement that one has, at some time or other, queried or challenged. No matter which answer is selected, cumulativeness is still absolutely crucial for understanding the *petitio*.

It still seems to us a primordial fact about the circularity inherent in the Woods-Walton dialogue-segment is that Bob grounds Wilma's acceptance of A on the basis of B at n+1, and then at n+5 grounds her acceptance of B on the basis of A. The critical factor here is what is called *groundedness* by Woods and Walton [226]. A dependency *petitio* occurs where for some statements A and B, A is grounded on B and B is grounded on A. And this mutual groundedness is what seems to occur in the dialogue-segment. It is because Mackenzie does not attempt to come to grips with the elusive and difficult notion of groundedness that his proposed solution to the problem of the Woods-Walton fragment bypasses the heart of the matter.

There also seems to us to be a basic distinction between (a) the *petitio*, and (b) replying to a challenge in a dialectical game with a statement already under challenge with respect to the replier. The case of (b) is characterized by the following form of dialogue-sequence in DC, which might be called *challenge busting*.

n Wilma: Why A?
n+1 Bob: B
n+2 Wilma: Why C?
n+3 Bob: A

Here Bob has broken a rule of DC by replying with A at n+4 even though he is under challenge with respect to A because of Wilma's move at n. This breach of Mackenzie's rules represents a shortcoming of rational strategy on Bob's part. What he has done is use as a basis for justification some statement to which Wilma, his opponent, is clearly not committed. Moreover, her lack of commitment is made signally evident by her challenge at n. Challenge busting is bad strategy generally, and even more so in a game where challenges are cumulative. Wilma can never possibly come around to accepting A in this sort of game.

But do we have arguing in a circle? Our answer is that it is not obvious that we have, so long as B and C are not in any evident way related by mutual groundedness in the argument-sequence in question. *Petitio*, in contrast to challenge busting, is paradigmatically represented by what we have been calling the *circle game*, represented in DC by the following sort of sequence.

n Wilma: Why A?
n+1 Bob: B
n+2 Wilma: Why B?
n+3 Bob: A

The above sequence represents a simple case. An even simpler case is as follows:

n Wilma: Why A?

Question-begging and Cumulativeness in Dialectical Games 265

n+1 Bob: A

But extended cases of the following form are possible.

n Wilma: Why A_0?
n+1 Bob: A_5
n+2 Wilma: Why A_1?
n+3 Bob: A_2

.   .

n+1 Wilma: Why A_{1-1}?
n+i+1 Bob: A_1
n+i+2 Wilma: Why A_1?
n+i+3 Bob: A_0

What is fallacious here? It would seem primarily to be that the sequence of justificatory responses loops back to the very point from which it started. Nothing is really gained.

$$A_0 \rightarrow A_1 \rightarrow A_2 \rightarrow A_{i-1} \rightarrow A_i \rightarrow A_0$$

And this is not merely challenge busting, but rather represents essentially what is characteristic of arguing in a circle. Each step finds a ground in its neighbour to the right. In DC, each move represents an attempt by Bob to find some statement which implies A_0 and which Wilma will accept. Up to A_1 the sequence seems to be running smoothly enough, but at the next step we know it is doomed to failure because we know that A_0 is the very statement we began with and Wilma does not accept it. The same sort of *petitio* would occur where Bob's move at n+i+3 consists of any statement, $A_0, A_1, ..., A_{i-1}, A_i$.

The interesting question is: what is the relationship between *petitio* and challenge busting? We now prove that *petitio* and challenge busting are not equivalent strategies in DC, because there can be a case of challenge busting that is not a case of *petitio*. On the other hand, every case of *petitio* is a case of challenge busting. Let us take the latter implication first. The characteristic of *petitio*, as we have seen above, is that there is a sequence of moves of the form R_{Chall}, that purport to come under the heading (iii) i.e. responding to the challenge by means of a statement not under challenge with respect to the responding participant,

$$A_0 \rightarrow A_1 \rightarrow A_2 \rightarrow ... \rightarrow A_i,$$

but that at some point an A_j is introduced into the sequence such that $A_j=A_0 \vee A_j=A_1 \vee ... \vee A_j=A_i$. In other words, at some point in the sequence there is a *looping back* to some previous point. But any point you care to choose in such a sequence, *say* A_j, represents a stage of the dialogue where A_j was posed in a challenge of the form ⌜Why A_j?⌝. Therefore every instance of *petitio* represents an attempt to reply at some stage with a statement that was under challenge at a previous stage with respect to the responding participant. Q.E.D.

We share with Aristotle a fondness for the epistemological distinction between circularity of demonstration and mere absence of evidence in demonstration.[9] The reader will by this point no doubt be aware that we have been led to make the precisely parallel distinction, now in dialectical terms, between circularity of argument and challenge busting in argument.

EPISTEMICALLY INCORRECT

(1) *Epistemic Inadequacy of Premiss*: choosing as premiss a proposition not better known than the inclusion.
(2) *Petitio Principii*: arguing that A is known on the basis that B and that B is known on the basis that A.

DIALECTICALLY INCORRECT

(1) *Dialectical Inadequacy of Premiss*: choosing as premiss a proposition that one's opponent is not committed to.
(2) *Begging the Question*: responding to ⌜Why A?⌝ with ⌜B⌝ and ⌜Why B?⌝ with ⌜A⌝.

In each case (2) is a special case of (1). Yet in neither case is (2) exactly the same failing of argument as (1).

Here we have differentiated the name of the fallacy of (2), according as it occurs in epistemic and dialectical contexts. However, the similarity is evident. In asking ⌜Why A?⌝ the participant in a formal disputation asks for evidential justification by appeal to some proposition, B. Similarly, in the epistemic case B provides the evidential justification basis for A. We have a fallacy in both cases because we have an argumental step of a binary sort from one proposition to another. When the step also loops back in the other direction we have a circular argument, regardless of whether the relation in question is dialectical or epistemic in character.

Thus we are led to the interesting insight that two questions must be separated:

(1) What *is* arguing in a circle (no matter the type of argument involved)?

(2) What is *wrong* with arguing in a circle?

The second question seems to call for separate answers in the epistemic and dialectical cases, whereas the first question may admit of a common answer for both. We conjecture that the difficulties for Hamblin and Mackenzie may stem from a failure to distinguish clearly between questions (1) and (2).

We now prove that there are sequences that do not conform to the above description of the *petitio* even though they do represent instances of challenge busting. Consider again the characteristic general form of challenge busting in DC above. To Wilma's ⌜Why A?⌝, Bob responds with B, then to Wilma's ⌜Why C?⌝ Bob responds with A. This represents two distinct sequences, A → B and C → A. Now it might seem that here we have the looping-back feature of the *petitio*. We started at A and finished with A. But we do *not* have another characteristic of the *petitio*, namely the sequential linking of justificatory steps into pairwise linkings. We stepped from A to B, and then from C to A, but there was no step from B to C. B and C are unrelated from the point of view of their groundedness in the dialogue sequence. Therefore there can be no *petitio* in this instance. In general then, an instance of challenge busting need not be a *petitio*, and it follows that the *petitio* is a special case of challenge busting in DC. This consequence accords with the thesis that challenge busting, while it may be bad argumentational strategy, need not always be fallacious in the way that *petitio* is.

4. BEGGING THE QUESTION IN DD

Mackenzie [127] investigates what happens in a system which, unlike DC, is *non-cumulative* with respect to challenges. The system DD resembles DC except that the commitment rule is changed so that ⌜Why P?⌝ is not included in its speaker's commitment store. What happens in the Woods-Walton dialogue fragment now? Interestingly enough, in DD Bob does not beg the question at n+5. So it would seem that cumulativeness of challenges is critical for the *petitio*. As Mackenzie notes [127, p. 10], Bob does not beg the question at n+5 precisely because Wilma had accepted A at n+2. Had she not done that, the dialogue would still have been illegal by R_{Chall}. By the non-cumulativeness of challenges, Wilma's move at n+2 effectively wiped out her commitment to ⌜Why A?⌝ and thereby affirmed her commitment to A, allowing a circle subsequently to occur. It becomes all the more apparent that cumulativeness is decisive for the analysis of the *petitio*.

Mackenzie however interprets the situation differently. He argues [127, p. 11] that Wilma's move is both unnecessary and unnatural, and that DD forestalls question-begging except in cases where the challenger acts in this unnecessary and unnatural way. The suggestions seems to be that the Woods-Walton segment is an artificial and unusual fragment of dialogue, in contrast to "normal" dialogues in DD, all of which later are spared the danger of circularity even without cumulativeness. In reply, it should be remembered that the Woods-Walton segment was constructed expressly for the "artificial" purpose of showing that a circle[10] can in principle be constructed in the non-cumulative Hamblin game (H). The question is one of the *theoretical capability* of certain classes of logical games to ban circles, and "normality" is not the issue.

True, Wilma's declaration of A is unnecessary, because she would already be committed to A by the failure to withdraw it at n+2 in (H), DC, or DD. And true, Hamblin [180, p. 267] does not in (H) give any examples of explicit concession under challenge. True again, Bob's remark at n+3 evidently "has nothing to do with the conversation" [127, p. 11]. All true, but they do not affect the main point, namely, that a *circle can occur in a system with non-cumulativeness of challenges*. Moreover, adding cumulativeness blocks the circle. Perhaps, ways could be found of adding rules to block the "unnatural" moves of the Woods-Walton fragment, so that an extension of DD without cumulativeness could be constructed so as to be free of circularity. But Mackenzie does not attempt to furnish a rule for this purpose, nor shall we, although it would be of some interest to do so. We would, however, offer one caveat. It is natural to suppose, or so we think, that the naturalness of an actual dialectical exchange will owe *something* to its propensity for evading vitiating fallaciousness. One must be careful, then, not to arrange theory in such a way that *perfect* dialectical naturalness is perfectly fallacy-free, by definition as it were. So the basic point remains - DD and similar non-cumulative games are circle-susceptible, and this is a crucial fact about them.

Circles *can* appear in DC and DD. In practical terms this means that a sophist can manipulate the rules to produce a *petitio* even if what he does may appear on reflection to be unnatural or even downright captious and devious. In theoretical terms, it means that there exists a certain characteristic class of games that permit a certain characteristic class of sequences of the Woods-Walton type to be circular if cumulativeness, whether of commitments or challenges, is waived, but that ban these sequences otherwise. So, we say again, that cumulativeness is crucial for blocking circles in formal dialogues.

5. ORGANIZING RETRACTIONS OF COMMITMENTS

We now venture to suggest that the problem posed by the Woods-Walton dialogue-segment is one of the organization of the retraction of commitments. This problem is quite general for dialectical theory, and it admits of numerous solutions, depending on the structure and purpose of the particular game at issue. Indeed such was an important factor implicit in the format of the medieval treatises on the Obligation game. We remember that in the medieval games, the strategy was to reject a *proposition* that is inconsistent with propositions one has already accepted. How, then, can one retain those commitments not inconsistent with the *propositum*, yet while giving up the ones that are inconsistent with it? Different commitment-sets may be available, and different principles may be formulated for choosing among these sets, depending upon the strategy that a disputant may have as his game-objective.

We offer one simple strategy that points the way to showing how dialectical games like (H), DC and DD can be extended so as to handle the Woods-Walton dialogue-segment. Basic to this solution is the conviction that in at least one clear sense an arguer should *not* be committed to all logical consequences of his commitments, but only to the "immediate" or "obvious" ones.

We begin with Hamblin's suggestion that for dialectical purposes a non-transitive relation of implication should replace the usual one. Hamblin introduces a relation *Imc* of *immediate consequence*. Three conditions are imposed [81, p. 144], but *Imc* is not otherwise specified:

(1) *Imc* implies the existence of a consequence relation between the set of statements denoted by its arguments, and, whenever there is a consequence relation between two sets of statements, there is a chain of immediate consequence relations leading from any set of locutions denoting the first to any set of locutions denoting the second.
(2) *Imc* is defined pairwise between single locutions in the much the usual way, except that we construe "single" locutions as "their" unit sets.
(3) If each of two locutions is an immediate consequence of the other, the locutions are said to be *dialectically equivalent*, and are dialectically everywhere interchangeable under all circumstances.

It is apparent from Hamblin's conditions (1), (2), and (3) that this consequence might be thought of as an instance of *relatedness implication*. Epstein [59] shows how the non-transitive relation of relatedness implication can be extended by the *transitive closure* of r called r_T, defined as follows.

$r_T(p,q)$ if, and only if, for some n, there are q_1, ..., q_n such that $r(p,q_1)$ and $r(q_1,q_2)$ and ... and $r(q_n,q)$.

A non-transitive r expresses the notion of *direct relatedness* whereas r_T expresses the idea of *indirect relatedness*. Hamblin's *Imc* may be viewed as a special case of relatedness implication and Hamblin's notion of dialectical consequence may be taken as indirect relatedness implication embedded in a dialectical system. Similarly, dialectical equivalence can be shown to be a special case of relatedness equivalence. A ↔ B (relatedness equivalence) is defined by Epstein [59] as (A → B) ∧ (B → A), where ∧ is classical conjunction and → is relatedness implication.

That part of relatedness logic which is a propositional calculus has a syntax composed of (1) propositions, P_0, P_1, P_2, ..., (2) negation,¬, (3) implication, →, (4) parentheses, (,), (5) meta-variables for wffs, A, B, C, ... A *model* is a truth valuation v: $\{P_i\}$ i ℘ → {T, F}, and a relatedness relation $r(P_i, P_j)$ which is a binary reflexive relation on $\{P_i\}$i ℘. Given r, R(A,B) is then defined to be the least relation of wffs which satisfies: (1) $R(P_i, P_j)$ ↔ $r(P_i, P_j)$ (2) R(A,¬B) ↔ R(A,B); R(¬B,A) ↔ R(B,A) (3) R(A,B → C) ↔ R(A,B) or R(A,C); R(B → C,A) ↔ (B,A) or (C,A). v is extended to a valuation V on all wffs via the following truth tables.

A	¬A
T	F
F	T

A	B	R(A,B)	A → B
any values		F	F
T	T	T	T
T	F	T	F
F	T	T	T
F	F	T	T

System R is comprised of sixteen axioms with a rule of detachment (A and A → B yields B), and is shown complete for its intended interpretation. An idea of how it works can be given by considering some classical theorems that fail. A → (B → A) fails because if A is not related to B, A could be true while B → A is false. [A → (B → C] → [B → (A → C)] fails because the antecedent can be true and the consequent false where A, B, and C are all true and A is related to B, B is related to C, and A is not related to C. Exportation fails. And "paradoxical" theorems such as ⌜¬A → (A → B)⌝ fail.

Question-begging and Cumulativeness in Dialectical Games 271

The philosophical applicability of relatedness logic to notions of entailment, and to informal fallacies as well, is discussed by Walton [200].

How are we to construe the concept of relatedness of statements in a dialectical game? This is not so much a technical question as one of understanding the motivation of the concept of relatedness as applied to dialectical argumentation. What does it mean to say that one statement is "related to" another in a dialectical setting? We believe that relatedness can best be understood in relation to the notion of implication that must be embedded in the rule for justificatory responses to a request of the form ⌜Why S?⌝. Let us take the game (H) as an instance. One allowable response to ⌜Why S?⌝ is the justificatory reply (d): ⌜Statements T, T ⊃ S⌝ for any T. Here Hamblin allows the classical ⊃ to stand in for the implication relation of (d). But, as we have seen, the unlimited transitivity of ⊃ and the attendant failure to distinguish between direct and indirect implication give rise to the problems of logical omniscience and call into question the realistic applicability of dialectical systems. It is here therefore, in a justificatory rule such as (d), that relatedness exhibits its utility. But if so, we might now well ask, what exactly could it mean here to say that S and T are "related"? The most satisfactory answer for a game like (H) is to say that T relatedly implies S if, and only if, S follows from T by a single application of one rule of inference of the dialectical game in question. That is, T directly implies S where there is a rule of the form Φ *to infer* ψ, for formulas Φ and ψ, and T is a substitution instance of Φ and S is a substitution instance of ψ. Then T indirectly implies S if, and only if, there is a series of applications of single rules that begins at T and ends at S, thereby establishing implication by transitive closure.

This notion of relatedness has 'bite' in the context of dialectical games of the sort Hamblin has in mind because we are thinking of a game in such a fashion that each participant has access to a common set of rules of inference for his use in responding to requests for the justification of statements. It is well, however, to leave open the possibility that relatedness might be applied in other ways to different sorts of dialectical contexts.

Given the establishment of a well-defined notion of immediate implication, a solution to our main problem is as follows. We can

(i) allow Bob and Wilma no "level of rationality" at all in the closure of their commitments, or
(ii) allow them the "limited rationality" of having their commitments closed under immediate implication.

Notice that in neither case are they "omniscient", and that we could allow one to be "more rational" than the other in a variety of ways; but we shall not explore these possibilities here. Now if we follow Hamblin's convention of not allowing ⌜Why A?⌝ if A is a commitment of the speaker, we have it that a *petitio* does or does not occur precisely as option (i) or (ii) is adopted. If (i) is adopted, the *petitio* occurs as outlined in section 2 and section 3. If (ii) is adopted then A and B must be commitments of both Wilma and Bob as well, because Wilma's commitment store contains B and ⌜if B then A⌝. By (ii), A must be included because it is an immediate consequence of B. Thus if (ii) is adopted, the move at n by Wilma is illegitimate and the Woods-Walton segment is blocked.[11]

Notes

Introduction

1. Douglas N. Walton [197a], and John Woods [218a].
2. Douglas N. Walton [191a]. See the review by Henry W. Johnstone Jr. *Noûs*, 21, 1987,69-72.
3. John Woods and Douglas Walton, [229], reprinted here as Chapter 8.
4. p. 271.
5. John Woods and Douglas Walton [224a].
6. Ralph H. Johnson and J. Anthony Blair [109a, p. 187].
7. Ibid., p. 189.

Chapter 1

1. The condition is a sham. Premisses may be omitted when they would be taken for granted, i.e., when in the relevant sense they could not be omitted.
2. Of course, if such predicates were admitted by our theory that would be tantamount to its claiming that the verbs of validity exhibit what Anthony Kenny in [113, ch. 7] calls "variable polyadicity", and thereupon the theory creates for itself the well-known difficulties of which Donald Davidson writes in "The Logical Form of Action Sentences" in [48].
3. Eike-Henner Kluge has drawn to our attention that this point has also been made by Gilbert Harman in [85, p. 43]
4. We may note that this fourth pass goes some way toward acknowledging, if not dispelling, some of our earlier hesitations: is the theory not relativistic? (Yes, and explicitly so.) Is there not a prejudice against argument predicate monadicity? (Yes, we are dealing with dialectic.) But what of the problem of variably polyadicity? (No comment.) And are we not committed only to a sham sense in which premisses may be omitted? (No comment.) And are we not still stuck with single-sources knowledge (or, now, acceptance?) (Apparently.)

5. Suppressing for now the relatively of 'good for v', 'acceptable to v', etc.

Chapter 2

1. As Aristotle recognized in Rhetoric, 1398, 18. [9]
2. The language of the evidence/expert testimony contrast is unfortunate for it may suggest, for example, that there can be no question of there being reliable expert testimony as to what in fact is the evidence, as in forensic medicine or criminology. Still, we shall suppose the intended distinction to be clear enough for our present purposes and shall not here seek an improved means of expressing it.
3. The diagonal proof is itself attended by a metamathematical controversy. The above example, therefore, is best understood as relative to those for whom diagonalization is not at issue.
4. This is not to say, of course, that the standard canons of confirmation could not be used to confirm that the intuition of an expert is on the whole reliable.
5. To avoid the very pitfalls they describe, 1.-6. have been directly quoted from De Morgan.
6. Not to be over-severe on occultism, let us understand that our occultist is a lucky and (very) minor fraud. His predictions lack a logos; he is privately as thunderstruck by his predictive record as everyone else.
7. Thus it would seem that even the gross logical form of a correct argument from authority is not quite captured by Salmon's statistical syllogism [167, p. 140-41].
8. See various well-known writings of Quine. It is worth noting that it is not necessary for the point at hand to accept Quine's account of theories as such; it suffices that some theories are underdetermined by the data and require extrapolation via the theorist's intuitive expertise.
9. Thus, by "intuitions" we mean pre-analytic judgments or beliefs not normally supported by appeal to explicit criteria.
10. The Delphi technique is not, of course, without its critics. See, for example, [184, ch. 5]. See the following as well [42], [43], [44], [27].
11. The interested reader may wish to consult some recent works by John Woods, e.g. [218].
12. The Delphi has other applications and has received its widest usage in technological forecasting [207, p. 2].

Chapter 3

1. Hamblin writes that 'to beg the question' is a reasonably accurate translation of Aristotle's expression 'το εν αρχη αιτεισθαι', meaning 'beg for that which is in the question-at-issue'. In the Aristotelian context of disputation, a person may ask to be granted certain premisses on which to construct his argument. The fallacy arises in asking to be granted the putative conclusion at issue. *Vide* C.L. Hamblin [80, p. 32f.].
2. *Vide* Hamblin [80]. Hamblin is virtually the only writer since Aristotle (with the possible exception of Alfred Sidgwick) seriously to confront the underlying theory of the informal fallacies in a systematic fashion. See our comments on Hamblin in [233].
3. Though, to be sure, the intuitive geometry of this paradigm is hardly that of the circle.
4. Irving Copi [40, p. 83]. Copi is an exemplar of this conception.
5. Where a premiss has the form 'P_1 & P_2 &... & P_k' ($k \geq 1$) and where P is the set $\{P_2, P_2, ..., P_k\}$ then a premiss-conjunct is any subset of P. In other words, a premiss-conjunct is simply a premiss or conjunctive component of a premiss. This way of putting it simplifies the statement of the conditions.
6. Thus if q is the conclusion and p the premiss, it would, by this condition, be circular for a to make the inference 'p, therefore q' provided the sentence 'p, but for all I know, not-q' would be indefensible for a to utter, in the sense of 'indefensible' of Hintikka's *Knowledge and Belief*, pp. 32-33. See below, note 13.
7. Since it might always be possible to come to know a premiss just by asking *someone* who is believed already to know, the condition comes to this: the premiss or premiss-conjunct must be unknowable unless someone at some time or other inferred it from the conclusion.
8. We could call (CI) De Morgan's Condition (see [50]). (CM) and (CP) are named for Mavrodes and Penelhum respectively.
9. Does this condition imply that knowledge is closed under equivalence? No indeed, and rightly. The point is that knowledge is *not* in general, closed under equivalence, and therefore that circularity attaches to those *non-general* cases where knowledge happens to be closed, or should be expected to be closed. But which cases are these?
10. Not a happy condition, even on its face. It is an unworkable condition in the absence of an account of nothing less than so-called propositional identity.
11. Morris Cohen and Ernest Nagel [37, p. 379]. Archbishop Whately quite clearly recognized these two distinct conceptions in stating that an argument is circular

In which one of the Premisses is either manifestly the same in sense with the Conclusion, or is actually proved from it, or is such as the persons you are addressing are not likely to know, or to admit, except as an inference from the conclusion [208, p. 220].

12. See A.R. Anderson and Nuel D. Belnap Jr. [2, p. 22]. But cf. John Woods [214, p. 315ff.].

13. 'Indefensible', in the sense of Hintikka. See [98, p. 32-3] *et passim*. Since inferences are reasoned expressions of belief or knowledge, the epistemic-doxastic vocabulary of 'indefensibility' is a fitter medium of analysis than the first order notion of 'inconsistency'. And for this reason, too, 'self-sustenance' (*idem.*) is preferable to 'validity', and 'virtual implication' [98, p. 57], to 'implication'.

14. For the sake of brevity, here and in the sequel we will use 'epistemic' generally to cover doxastic and acceptance settings as well.

15. Cf. Aristotle, *Prior Analytics*, 64b 33.

16. But are not the rules of entailment also rules of inference? No see John Woods and Douglas Walton [233], John Woods [212], and Gilbert Harman [85].

17. For an argument to the essential subjectivity of circularity see Davis Sanford [167].

18. See John Woods and Douglas Walton [233] and John Woods [212].

19. If the legalistic barbarism may be forgiven.

20. Not to say necessarily a fallacious involvement. See John Woods and Douglas Walton [225].

21. Jaakko Hintikka, 'Are Logical Truths Analytic?', *Philosophical Review*, (1965), 178-203, Hintikka, "Knowing Oneself" and Other Problems in Epistemic Logic' *Theoria* 32 (1966), 1-13. Hintikka, 'An Analysis of Analyticity', in *Deskription, Analytizität und Existenz*, Verlag Anton Pustet, Salzburg und München, 1966, 193-214. Hintikka, 'Are Logical Truths Tautologies?', *ibid.*, 215-233. Hintikka, 'Kant Vindicated', *ibid.*, 234-253. Hintikka, 'Kant and the Tradition of Analysis', *ibid.*, 254-272. Hintikka, 'Are Mathematical Truths Synthetic A Priori', *The Journal of Philosophy* 65 (1968), 640-651. Hintikka, Information, Deduction and the A Priori', *Noûs* 4 (1970), 135-152. Hintikka, 'Knowledge, Belief and Logical Consequence', *Ajatus* 32 (1970), 32-47. Hintikka, 'Surface Information and Depth Information', in *Information and Inference* (ed. by Jaakko Hintikka and Patrick Suppes), D. Reidel Publ. Co., Dordrecht, 1970, 263-297.

22. Aristotle writes in the *Topics* (162b 32) that one way of begging the question "... occurs whenever are one begs universally something which he has to demonstrate in a particular case ...."

23. For simplicity we shall assume that arguments in this form are put forward and received without any discord arising from their enthymematic characters.
24. See Bertrand Russell [165, p. xxi-xxii]. Cf. W.V. Quine [155, p. 116f.]
25. Oliver Herford's *Alphabet for Celebrities* provides an example that is both illustrative and amusing.

> Q is for Queen, so noble and free;
> For further particulars look under V.
>
> ....
> V is Victoria Noble and true;
> For further particulars look under Q;

Or:
> A number is either the number zero or the successor of some number.

26. Nor do we deal here with the circular aspects of the Complex Question.
27. As in the setting, "But that definition I do not accept; it is a bad definition".
28. As in the setting, "I hadn't realized that this was the definition of 'cause'; well then, I accept what you say".
29. Rescher exemplifies this kind of intuition in writing it is clear that circular arguments need not be incorrect. For example, the valid, but trivial argument

> Rome is the capital of Italy.
> Therefore: Rome is the capital of Italy.

is defective only in that it is wholly without persuasive force. Anyone in doubt about the conclusion must necessarily also question the premiss. Nicholas Rescher [159, p. 85].
30. We would like it to be clear that we are *not* saying that effectiveness is by itself a sufficient condition of a good argument.

Chapter 4

1. Much the same conclusion is wanted for the same question concerning the *ad populum* and the *ad misericordiam*, but we do not pursue these matters here.
2. It seems to follow that there can be no such thing as an *equivocal argument*, and thence that equivocation is not a fallacy. In other words, the requirement above ignores Aristotle's fallacies within language.

3. Not that *belief* falls outside the scope of logic entirely - see [79]. It's just that it's hard to see how *force* as a cause of belief comes within the scope of logic.

Chapter 5

1. See Hamblin [80, p. 161].
2. See Woods and Walton [233], and [234].
3. See [63]. This passage is also quoted by Hamblin [63], [80].
4. We will not embark on the project on constructing a logic of assertion here, but only mention that we have in mind a performative notion such as those studied in Hintikka [95] and Walton [198]. We would add that assertion might also be profitably studied in the dialectical framework of Hamblin [80, ch. 8], rather than what we would call an exclusively monolectical framework; see [233].
5. Much interesting background to the *ad hominem* from a point of view of rhetoric is provided in [143].
6. See Hamblin [80, p. 207]
7. The Obligation Game originates with the ancient Greek debates, but one of the earliest treatises on it is ascribed to William of Sherwood. Hamblin gives a two-person version based on a finite propositional language. The object of the game is to trap the opponent in inconsistency in a finite number of moves.
8. The Kripke semantics posits points on a tree-model as representing "advancing states of knowledge". For a brief outline and discussion of this semantics in the context of informal fallacies, see our paper [226].
9. As before, p must be a statement out of an object-language. Thus the Liar is disqualified as a *tu quoque*, as are other deviances of semantically closed languages.
10. A central notion of (TQ3) is that of *bringing something about*. For attempts at its logical analysis, the reader might wish to consult Pörn [150], [149]. See also Åqvist [6] and Walton [194] and [197].
11. For some discussion, see Rescher and Helmer [163]. A similar problem is presented by the *ad baculum*. See [235].
12. See Helmer [89].
13. See Woods and Walton [225], for further discussion.
14. Salmon [167] uses the term 'anti-authority,' but we wish to carry over the distinction [167] between authority and a special case of it, expertise.
15. A little more finely put, the possible rejections, in increasing order of severity, would seem to be these. (i) X can't claim to know that s on the basis of p, q and r; (ii) X can't claim to know

s; (iii) s, which X concludes, is false. The likelihood of *ad ignorantiam* varies proportionately with the degree of severity of these rejections.
16. Both Gardner [69] and Salmon [167], following Gardner, set out criteria for crankhood. Our set of criteria is based on theirs, although not identical with either account.
17. Some would say that Wilhelm Reich was a crank. We use his case as an example without necessarily accepting the attribution.
18. See Gardner [69].

Chapter 6

1. Though there is no doubt that the point (or something very like it) is adumbrated by Aristotle in the *Posterior Analytics*, 73a 10.
2. Mill is not mentioned by name. De Morgan concedes that there is "much ingenuity" in the argument, but he does not credit it to a specific source.
3. Of course, whether the resulting argument is single-premissed will depend on which superfluous premisses are deleted. For example, argument (5) has two non-superfluous premisses when its first premiss has been deleted.
4. Not to overlook the welcome and useful efforts of C.L. Hamblin to reverse this trend. See Hamblin [80].

Chapter 8

1. The resolution of this example raises a number of vexing and important problems for the explication of concepts of relative possibility and power-attribution. See Lehrer and Taylor [122], Hilpinen [90] and Walton [203]. At 177b 23ff, Aristotle suggests, intriguingly if obscurely, that the difference we seek may reside in the observation that it is not the same thing 'to do a thing in the way he can' and 'to do it in every way in which he can.'
2. The original from Peter does actually follow Aristotle in using 'and'. In [144] 24va 30, we find the expressions *duo et tria sunt quinque (composita) and duo quinque et tria sunt quinque (divisa)*.
3. This argument is, however, clearly meant by Peter to be disjunctive. In [144] 27va 20, we find:

Omne animal est rationale vel irrationale.
Sed non omne animal est rationale.
Ergo omne animal est irrationale.

4. Quine suggests, [154, p. 189] that the traditional distinction between distributive and collective predication is a tentative form of the set-theoretical distinction between ε and , membership and inclusion. Quine cites the mediaeval paralogism: Peter is an apostle, the apostles are twelve, therefore Peter is twelve. According to our analysis, it would be quite possible for a fallacy of composition (of our second kind) to occur through confusion of ε and $\supset$, but we do not think that the paralogism cited is a case of this type.

4a While clearly a welcome distinction for those who take composition and division to be fallacy-capable, difficulties nevertheless lie in wait. For consider a chain each link of which is a bracelet, and let each bracelet in turn be made of interlocking rings. Burge's theory precludes our saying that the rings are components of the chain, a consequence that not everyone would accept.

5. For related peculiarities of conjunction and whether they are peculiarities rightly attributed to 'and,' see Woods [213] and Gahringer [67].

6. A perspective that non-philosophers can also share. Thus a contemporary economist:

> The study of economic problems can be fascinating. However, ... there are several pitfalls... These pitfalls can be listed under the following headings: preconceptions; self-interest; problems of definition; fallacy of composition; and false analogy. Archer [7] (emphasis ours).

Archer continues.

> By fallacy of composition we mean the mistake of assuming that what is true for part of a group must necessarily be true for the group as a whole. Thus, whereas an individual farmer may be better off by his increasing production, farmers as a whole, may be worse off. Archer [7, p. 46].

7. This, or something like it, is clearly the view of the good doctor Watts [206, p. 320-1]:

> The *Sophism of Composition* is when we infer any Thing concerning ideas in a *compounded Sense*, which is only true in a *divided Sense*.

There follow examples, then a similar definition of Division with examples, after which:

> This sort of Sophism is committed when the Word *All* is taken in a *collective* and a *distributive* Sense, without a due Distinction, ... It is the same *Fallacy* when the universal Word *All* or No refers to Species in one proposition, and to Individuals in another;

Watt's examples span both the Mediaeval and modern conceptions on the fallacy. Thus:

> And when it is said in the Gospel that *Christ made the Blind to see*, and the *Deaf to hear*, and the *Lame to walk*, we ought not to infer here that Christ performed Contradictions;

and

> All animals were in Noah's Ark, therefore, No animals perished in the Flood.

Chapter 9

1. The reader is referred to our article [225] for remarks on the structure of explanation appropriate to the statistical, causal, and authority-based informal fallacies.
2. True, in the *Treatise* Hume speaks of cause as "an object precedent contiguous to another, and so united with it, that the idea of the one determines the mind to form the idea of the other, and the impression of one to form a more lively idea of the other" [106a, p. 170, italics added]. However, we share the view that Hume is not here giving truth conditions of "x caused y", but rather of "A believes that x caused y". It is true, of course that one looks in vain in the Treatise for crisp and entirely satisfactory evidence that Hume intended or recognized this distinction.
3. See [106a, p. 73 ff].
4. With a view to greater exactitude than we here venture, Mackie has characterized a cause as a condition that is "an *insufficient* but *necessary* part of a condition which is itself unnecessary but sufficient for the result" [128], [129, p. 62]. Thus the acronym, "INUS condition". Suppes has given a set-theoretic definition of this concept in terms of probabilities. In general, Suppes has worked out various minimal probabilistic causal algebras that go some way towards meeting the general conditions we have laid

down as characterizing the notion of cause, and the reader who desires a more specific account of the causal relation might consult this excellent work [180].

5. We will not attempt to isolate these "Pragmatic" factors here. The sort of parameters we have in mind are those discussed at length by H.L.A. Hart and A.M. Honoré in [87]. Gasking is also helpful in this connection, [71].

6. Another interesting approach is that of David Lewis, see [124].

7. Relatedness semantics arose through the contributions of a number of participants in the *Logic Seminar* of Victoria University of Wellington and might best be called "Victoria Semantics" or "the Victoria Group Semantics". See Richard L. Epstein, *The Semantic Foundations of Logic*, vol.1 (*Proposional Logics*), Dordrecht, Kluwer, to appear.

8. But see also Suppes' demurrer, [180, p. 5].

9. Something like this claim often appears in introductory logic texts. Hamblin cites one [80, p. 37], but we shall not attempt to make a list of likely culprits, since the attractiveness of the notion is obvious enough.

10. See the discussion in Hamblin, p. 37f. We might add here that care is needed in making general pronouncements of this kind about causation and correlation. For example, it might seem reasonable to hazard the conjecture that positive correlation always has some causal evidential power, other factors being held in abeyance. But consider the positive correlation between the stork population and (human) birth rate in an English village. Most of us would think it unlikely that this correlation indicates or even suggests a causal connection. See also note 14.

11. We waive any claim that these five premisses constitute a sufficient condition for inductive correctness of this kind of causal argument. Our thesis is that each of them is a necessary condition for correctness of this form of argument.

12. For an explanation of these terms, see [187, p. 176 ff].

13. Providing the causal relation is transitive. Suppes' discussion shows that the assumption of transitivity is questionable. Refusal to grant the assumption will require some modification of the view of intervening causes proposed above. See [30, 3.3].

14. Richard Epstein points out, however, that the storks case may be susceptible to analysis as violating (P3). See the discussion in [32, p. 173 ff.].

15. Lewis' approach is also based on modal logic, but makes use of notion of comparative similarity that takes it beyond the framework of Burks.

16. See [221]. Our relation of befortification is intriguingly close to Suppes' concept of *prima facie* cause. The analogy is brought

out particularly clearly by the discussion in Domotor, [52, p. 14 ff].
17. In a talk given to the *Logic Seminar*, July 18, 1975, at Victoria University of Wellington, New Zealand.
18. We should like to thank Professor Arthur Burks for furnishing us with a manuscript copy of his book, *Cause, Chance and Reason* [30]. We are grateful to David Loewen and Norman Swartz for comments, and Richard Epstein and Robert Goldblatt for comments, and Richard Epstein and Robert Goldblatt for helpful discussions of certain troublesome points.

Chapter 10

1. See [102a].
2. See [117, p. 97ff.].
3. See [117, p. 99].
4. See [117, p. 99 and 94].

Chapter 11

1. Though not by Professor Popper.
2. For some background to the notions developed in this section, vide [33, ch. 2], also 'Preface to the Second Edition.' Discussion of some applications of inductive logic to the domain of informal fallacies is included in John Woods and Douglas Walton, [221].
3. For background to the notions developed in this section, vide Jaakko Hintikka, [98] and Kathleen Johnson Wu, [107] also the review of the latter by Douglas Walton [195]. For other applications of epistemic logics to the informal fallacies, see John Woods and Douglas Walton, [234] and Douglas Walton [197].
4. See also the comments in Hamblin. [80, p. 17ff.]

Chapter 12

1. In Geach, [73, p. 165] the phrase "by means of logic" is changed to "by *a priori* methods".
2. Intuitionists may be the exception. Kripke's semantics for intuitionistic logic [117] may fairly admit of the adjective 'epistemic'. Whether Kripke might identify entailment with the intuitionistic calculus is of course another matter, but the point is that some intuitionists may be ready to make this identification.
3. For more on this, see also Hamblin [80].

4. De Morgan in his *Formal Logic* defended a variant of this conception. See Douglas Walton, "Mill and De Morgan on Whether the Syllogism is a *Petitio*" (summary) [196].
5. See also the discussion of disjunctive syllogism in Woods and Walton [234].
6. Theorems in some system or other. We concede with von Wright that the phrase "means of logic" is vague, see [190, p. 181]. But a vague definition is not a bad one if its vagueness matches the vagueness of its *definiendum*. As von Wright notes [190, p. 183], the definition allows for a pluralistic interpretation of this phrase as "by *some* means of logic".
7. This way of proceeding fits in very nicely with von Wright's remarks in [190, p. 185]: "Truth which has become established 'by means of logic', is, moreover (logically) necessary truth".
8. By which we, and we think von Wright, mean that concept (or concepts) of possibility, namely logical possibility, that T, S4, S5, and the other standard systems of modal logic are often said to represent. It has been conjectured by Lemmon (see Hughes and Cresswell [105, p. 80]), that S5 is the most adequate system to explicate this sense of possibility.
9. Rather it is a matter of "applied logic", or perhaps "informal logic".

Chapter 13

1. See also our paper, [231] where we argue that Mill's adherence to an essentially epistemic conception of proof is not, as we are often told, evidence of that offensive nineteenth-century psychologism of which logic needed to free itself, but rather a legitimate investigation on the interface of mathematics and psychology that is vindicated by some current developments.

Chapter 15

1. [50, p. 242]; *Finis rei est illius perfectio; mors est finis vitae; ergo mors est vitae perfectio.*
2. This example is similar to, but not identical with one given by [208, p. 195].
3. Our application of the concept of cognitive dissonance to the fallacy of equivocation is, as far as we know, a novelty. But studies on cognitive dissonance in the context of the psychology of argumentation, communication and persuasion have been carried out. See for example [103].

Chapter 16

1. These notions are more fully developed in Hamblin [80].
2. To be sure, a universal, personal appeal to "everyone" is part of the premisses, but this appeal need not involve any sort of enthusiasm and therefore lacks an emotive element of the *ad populum*.
3. This theme is further developed in [225].
4. The notion of subject-matter overlap is due to David Lewis.
5. A complete relatedness calculus is developed by Richard L. Epstein [59].
6. Further elaboration of these points is made in Walton [200].
7. Or at any rate, it should be as formal as possible.
8. For more on this property, see [234].

Chapter 17

1. Saul Kripke, [117] and Jaakko Hintikka [101]. To be clear about this point, it is not in fact the case, so far as we know, that there exists any theory of circularity that qualifies for the status of a mathematical system. Fragments of such a theory exist which involve the use of Kripke's intuitionistic logic (which is a mathematical system) and also Hintikka's dialogical system (which is not a mathematical system).
2. Hao Wang [205a].
3. Wang, ibid.
4. The interested reader might wish to consult some of the following: [225], [233], [234], [222], [236], [235], [226], [221], [232], [227], [228], [238], [243], and [223].
5. See [226, p. 82-9].
6. See [236, p. 574-7, 591-3].
7. Richard Routley [163a]. In this paper Routley develops a system in which contradictions can be true yet in which not everything is true. In other words, in such a system, contradictions do not make for theoretical psychosis. Routley furnishes an argument intended to show that the world is in fact inconsistent. This metaphysical claim not only provides the motivation for a logic in which such contradictions might obtain, but it is also of very direct relevance and importance for any analysis of the kind of *ad hominem* argument in which there is a charge of inconsistency.
8. See, for example, [16], [6], [86a], [209a], and [202].

286 *Notes*

9. For a discussion of the distinction between inference and entailment rules the reader could consult Woods and Walton, [234, p. 112-3, 125-6], see also [233].
10. See also the Appendix to the present paper [228, p. 388-96].
11. See for example [221], [233], [234], and [226].
12. Waiving for now discussion of an interesting thesis that one tends to associate with Hilary Putnam, namely, that logical theories are also empirical.
13. Of course, various bits and pieces may yield up quite convincing axiomatizations, but that is a different point.
14. We have appropriated most of this list from Baruch A. Brody [254, p. 64].
15. See especially [181, p. 392-95].
16. For a more complete examination of the suitability of the formal theory of aggregates to the analysing of composition and division, the reader may wish to consult [228].

Chapter 18

1. The traditional question is actually 'Have you stopped beating your wife?' but since recent surveys indicate that husband-beating is also widespread, it seems only fair to allow for that possibility.
2. Belnap and Steel say the same thing.
3. This is to acquiesce, though not too heavily, in the traditional doctrine that fallacies are invalid arguments that *seem* valid. Thus any fallacy will have two sides to its counterfeit coinage - a formal side and a more pragmatic side. This does not mean, of course, that every instance of an informal fallacy must always seem valid at some time to everyone, or something of the sort. Merely that it must be a form of argument that is of some general interest in studying patterns of rational disputation between participants in argument. For more comments on this dichotomy see [234].
4. For elaboration on this theme see [234].
5. Not that fallacies cannot be studied from a rhetorical point of view.
6. See also the useful discussion in [152], and [208].
7. Interesting discussions of errors in dealing with conjunctive propositions and questions are to be found in [72, p. 180-20] and [74, p. 77f.]. Professor Geach tells us that he thinks no fallacy is committed by the mere asking of a question of the form Is it the case that both p and q?'. We agree with this thesis and hope that everything we say is consistent with it. However, as Geach shows, a formal fallacy occurs on the part of such a questioner if he

infers ¬p ∧ ¬q from a negative answer to this question: the inference from ¬(p ∧ q) to ¬p ∧ ¬q is formally invalid.
8. It is interesting to remark that the Black and White Fallacy has an element of "forcing" that we saw to be characteristic of Many Questions. This element is made especially clear in the treatment of Aristotle's example by William of Sherwood [210]:

> Suppose that two things are pointed to, one of which is good, the other bad. 'Are these things good or not good?' If one takes the affirmative (*si concedat*), one is necessarily refuted, for it follows that what is not good is good. But if one takes the negative (neget), refutation seems to follow although it does not follow; for this does not follow: 'they are not good; therefore this one of them is not good.'

Here the question is actually safe, the presupposition being $G \vee \neg G$ where G represents the statement 'These things are good.' Here also the fallacy is one that pertains essentially to conjunction. Just as ¬(A ∧ B) does not imply ¬A, so 'x and y are not (both) good' does not imply that x is not good. But the essential point to note here is that, as William makes clear, the trap posed by the question consists in its attempt to force the answerer to choose among two alternatives, each of which is untenable, to force the answerer to the acceptance of a propositions that is unwelcome to him. The Black and White Fallacy shares some characteristics with the Fallacy of Many Questions.

Chapter 19

1. Green and Hamblin [78] outline interesting variants of this game, but the basic structure is similar. A sequence of moves according to a set of rules is continued until the opponent says *Cedat tempus* [Time's up!], at which point the opponent wins if the respondent has been trapped in inconsistency or the respondent wins if he has avoided inconsistency up to that point.
2. Thus what in effect corresponds to the property of cumulativeness in Kripke models for intuitionistic logic is taken to be an optional element in constructing games of dialogue. See Woods and Walton [226, p. 20].
3. Here is such a structure: For the reader's convenience we set out the initial commitment-store of each participant in brackets at the head of the tableau. A superscript indicates at which step an addition is made; a stroke indicates deletions; and a superscript at the head of the stroke marks the step at which that statement was removed from the store.

VII

WHITE [A ⊃ B, B ⊃ A, A², B³]	BLACK [A, B, A ⊃ B, B ⊃ A, C]
1. Why A?	Statements B, B ⊃ A
2. Statement A	Statement C
3. No commitment B; why B?	Statements A, A ⊃ B
4. Statement B	

Is this circular? For pros and con A, see Woods and Walton [226, p. 79-83].

4. This asymmetry, by the way, makes Rescher's conception of dialectic somewhat narrower than Hamblin's, for according to Hamblin a more symmetrical interpretation of the roles of the participants is possible. See Hamblin [80, p. 265].

5. Just as *Imc* is critical in the structures proposed by Hamblin. See Hamblin [81, p. 144], and below section 5.

6. An *equivalence petitio* occurs when the conclusion is equivalent to one of the premisses. See Woods and Walton [285, p. 107-8].

7. A *dependency petitio* occurs when the acceptance of a premiss requires acceptance of the conclusion. See Woods and Walton, *idem*.

8. Relatedness implication is explored in Epstein [89], and its applications to philosophical logic are developed in Walton [200].

9. This theme is elaborated in our article, to appear.

10. Or anyhow a pattern that gives a strong impression of circularity.

11. We are assuming here that conventions of the cumulativeness of challenges are not in effect; but much of the same strategy as above is applicable whether it is cumulativeness of challenges or commitments that are at stake.

Bibliography

1. Abate, Charles J.
   1979 'Fallaciousness and Invalidity'. *Philosophy and Rhetoric* 12, 262-66.
2. Anderson, A.R. and Nuel D. Belnap Jr.
   1962 'Tautological Entailments'. *Philosophical Studies* 13, 9-24.
3. Anderson, A.R. and N.D. Belnap
   1962 'The Pure Calculus of Entailment'. *The Journal of Symbolic Logic* 27, 19-52.
4. Anderson, John
   1938 'The Problem of Causality'. *Australasian Journal of Philosophy* II, 127-142.
5. Anderson, K.
   1971 *Persuasion: Theory and Practice.* Boston: Allyn and Bacon.
6. Åqvist, Lennart
   1965 'A New Approach to the Logical Theory of Interrogatives'. *Filosofiska Studier* Uppsala.
7. Archer, M.
   1973 *Introductory Macroeconomics: A Canadian Analysis.* Toronto: Macmillan.
8. Aristotle
   1928 *De Sophisticis Elenchis. The Works of Aristotle Translated into English* 1, ed. by W.D. Ross, Oxford University Press, London.
9. Aristotle
   1928 *The Works of Aristotle.* Ed. by W.D. Ross, Oxford: Clarendon Press.
10. Aristotle
    1958 *Topica et Sophistici Elenchi* translated by W.A. Pickard-Cambridge, ed. by W.D. Ross, Oxford: Clarendon Press.
11. Arnauld, Antoine
    1964 *The Art of Thinking* translated by James Dickoff and Patricia James. New York: Bobbs-Merrill.

12. Bar-Hillel, Yehoshua
    1964 'More on the Fallacy of Composition'. *Mind* 73, 125-126.
13. Barker, John A.
    1976 'The Fallacy of Begging the Question'. *Dialogue* XV, 241-55.
14. Bartley, W.W. (III)
    1964 'Rationality Versus the Theory of Rationality'. Ed. Mario Bunge, *The Critical Approach to Science and Philosophy*. London: Free Press of Glencoe, 3-31.
15. Belnap, Jr. Nuel D.
    1963 *An Analysis of Questions: Preliminary Report* Santa Monica, System Development Corporation.
16. Belnap, Nuel D. Jr. and Thomas B. Steel
    1976 *The Logic of Questions and Answers*. New Haven and London, Yale University Press.
17. Bem, D.
    1970 *Beliefs, Attitudes, and Human Affairs* Belmont, California: Wadsworth.
18. Bennett, Jonathan
    1969 'Entailment'. *Philosophical Review* 78, 197-236.
19. Bentham, Jeremy
    1962 *The Works of Jeremy Bentham* (ed. by John Bowring), New York: Russell and Russell, 375-487.
20. Berger, George
    1973 'Temporally Symmetric Causal Relations in Minkowski Space-Time'. *Space, Time and Geometry* (ed. by Patrick Suppes), Dordrecht, Reidel, 56-71.
21. Blalock Jr., H.M.
    1961 'Correlation and Causality: The Multivariate Case'. *Social Forces* XXXIX, 246-251.
22. Blanshard, Brand
    1962 *Reason and Analysis*. LaSalle, Illinois: Open Court.
23. Bonsanquet, Bernard
    1928 *The Essentials of Logic*. London: Macmillan & Co.
24. Bochenski, I.B.
    1970 *A History of Formal Logic*. trans. and ed. by Ivo Thomas, Notre Dame, Indiana: University of Notre Dame Press.
25. Bradley, F.H.
    1922 *The Principles of Logic* I, London: Oxford University Press.
25a. Brody, Baruch A.
    1976 'Logical Terms, Glossary of', *The Encyclopedia of Philosophy*, 5-6, New York: Macmillan Publishing Co. and The Free Press.

26. Broyles, James E.
    1975 'The Fallacies of Composition and Division'. *Philosophy and Rhetoric* 8, 108-113.
27. Brown, Bernice B.
    1968 *Delphi Process: A Methodology Used for the Elicitation of Opinions of Experts*. Rand Corporation, 3925.
28. Burge, T.
    1977 'A Theory of Aggregates'. *Noûs* 11, 97-118.
29. Burge, T.
    1975 'Truth and Singular Terms'. *Noûs* 8, 309-326.
30. Burks, Arthur
    1977 *Cause, Chance, and Reason*, University of Chicago Press.
31. Burks, Arthur
    1951 'The Logic of Causal Propositions'. *Mind* LX, 363-382.
32. Campbell, Stephen K.
    1974 *Flaws and Fallacies in Statistical Thinking*. Englewood Cliffs: Prentice-Hall.
33. Carnap, Rudolf
    1962 *Logical Foundation of Probability*. University of Chicago Press.
34. Caroll, Lewis
    1895 'What the Tortoise said to Achilles'. *Mind* 4, 278-280.
35. Chisholm, Roderick
    1954-55 'Law Statements and Counterfactual Inference'. *Analysis* XV, 97-105.
36. Chomsky, Noam
    1967 *Current Issues in Linguistic Theory*. The Hague: Mouton.
37. Cohen, Morris R. and Ernest Nagel
    1934 *An Introduction to Logic and Scientific Method*. New York: Harcourt, Brace & World.
38. Colby, Kenneth Mark
    1968 'A Programmable Theory of Cognition and Affect in Individual Personal Belief Systems'. *Theories of Cognitive Consistency: A Sourcebook*. Rand McNalley, Chicago, 520-525.
39. Cole, R.
    1965 'A Note on Informal Fallacies'. *Mind* 74, 432-433.
40. Copi, Irving M.
    1972 *Introduction to Logic*, 4th ed., New York, MacMillan.
41. Cox, Richard T.
    1961 *The Algebra of Probable Inference*. Baltimore: The John Hopkins Press.

42. Dalkey, Norman C.
    1969 *The Delphi Method: An Experimental Study of Group Opinion.* The Rand Corporation, RM-5888-PR.
43. Dalkey, N., B. Brown and S. Cochran
    1969 *The Delphi Method II: Structure of Experiments.* RM-5957-PR.
44. Dalkey, N., B. Brown and S. Cochran
    1969 *The Delphi Method III: Use of Self Ratings to Improve Group Estimates.* RM-6115-PR.
45. Dalkey, N., B. Brown and S. Cochran
    1970 *The Delphi Method IV: Effects of Percentile Feedback and Feed-In of Relevant Facts.* RM-6118-PR.
46. D'Angelo, E.
    1970 'Philosophers as Critical Thinking Consultants'. *School and Society* 88.
47. D'Angelo, E.
    1971 *The Teaching of Critical Thinking.* Amsterdam: B.R. Gruner.
48. Davidson, Donald
    1967 'The Logical Form of Action Sentences'. *The Logic of Decision and Action.* Pittsburgh: University of Pittsburgh Press.
49. Deegan, John Jr.
    1974 'Specification Error in Causal Models'. *Social Science Research* III, 235-259.
50. De Morgan, Augustus
    1847 *Formal Logic* London: Taylor and Walton.
51. Dewey, John
    1938 *Logic: The Theory of Inquiry.* New York: Henry Holt and Co.
52. Domoter, Zoltan
    1973 'Causal Models and Space-Time Geometries'. *Space, Time and Geometry* ed. Patrick Suppes. Dordrecht; Reidel. 1-55.
53. Ducasse, C.J.
    1966 'Critique of Hume's Conception of Causality'. *Journal of Philosophy* LXIII, 141-148.
54. Eberle, Rolf
    1974 'A Logic of Believing, Knowing, and Inferring'. *Synthese* 26, 356-382.
55. Ehninger, Douglas
    1974 *Influence, Belief, and Argument.* Glenview, Illinois: Scott, Foresman and Co.
56. Ellul, Jacques
    1972 *Propaganda.* New York; Knopf.

57. Ennis, R.
    1969 *Logic in Teaching*. Englewood Cliffs, N.J.: Prentice-Hall.
58. Ennis, R.
    1969 *Ordinary Logic*. Englewood Cliffs, N.J.: Prentice-Hall.
59. Epstein, Richard L.
    1979 'Relatedness and Implication'. *Philosophical Studies*.
60. Ewing, A.C.
    1951 *The Fundamental Questions of Philosophy*. New York: The MacMillan Co.
61. Fearnside, W., and W. Holther.
    1959 *Fallacy: The Counterfeit of Argument*. Englewood Cliffs: Prentice-Hall.
62. Festinger, Leon
    1957 *A Theory of Cognitive Dissonance*. Roe, Peterson.
63. Finocchiaro, Maurice A.
    1974 'The Concept of *Ad Hominem* Argument in Galileo and Locke'. *The Philosophical Forum* 5, 394-404.
64. Fischer, David Hackett
    1970 *Historian's Fallacies*. New York: Harper & Row.
65. Fisk, Milton
    1974 *Nature and Necessity: An Essay in Physical Ontology* Bloomington. Indiana: Indiana University Press.
66. Fitch, F.B.
    1963 'A Logical Analysis of Some Value Concepts'. *Journal of Symbolic Logic* 28, 135-142.
67. Gahringer, R.E.
    1970 'Intensional Conjunction'. *Mind* 79, N.S.
68. Gailei, Galileo
    1914 *Dialogues Concerning Two New Sciences*. translated by Henry Crew and Alfonzo de Salvio. New York: Dover.
69. Gardner, Martin
    1952 *Fads and Fallacies in the Name of Science*. New York: Dover.
70. Garrett, L.
    1960 *Philosophy in High School*. Jacksonville, Ill.: MacMurray College.
71. Gasking, Douglas
    1955 'Causation and Recipes'. *Mind* LXIV, 479-487.
72. Geach, P. T.
    1972 *Logic Matters*. Oxford, Blackwell.
73. Geach, P. T.
    1958 'Entailment' *Proceedings of the Aristotelian Society*, supplementary vol. 32, 157-172.
74. Geach, P. T.
    1976 *Reason and Argument*. Oxford: Blackwell.

75. Gerber, D.
    1974 'On *Argumentum ad Hominem*'. *The Personalist* 55, 23-29.
76. Good, I.J.
    1961 'A Causal Calculus'. *The British Journal for the Philosophy of Science* Part I, XI., 305-318 and Part II, XII. (1962), 3-52.
77. Goodman, Nelson
    1955 *Fact, Fiction and Forecast*. Cambridge, Massachusetts: Harvard University Press.
78. Green, Romuald, and C.L. Hamblin
    William of Sherwood on Obligation, unpublished.
79. Grofman, Bernard and Gerald Hyman
    1973 'Probability and Logic in Belief Systems'. *Theory and Decision* 4, 179-195.
80. Hamblin, C.L.
    1970 *Fallacies*. London: Methuen.
81. Hamblin, C.L.
    1971 'Mathematical Models of Dialogue'. *Theoria* 37, 130-155.
82. Hanser, Richard
    1970 *Putch!* Pyramid Books, New York.
83. Harary, Frank
    1969 *Graph Theory*. London: Addison-Wesley.
84. Harary, Frank, Robert Z. Norman and Dorwin Cartwright
    1965 *Structural Models: An Introduction to the Theory of Directed Graphs*. New York: Wiley.
85. Harman, Gilbert
    1970 'Introduction: A discussion of the Relevance of the Theory of Knowledge to the Theory of Induction'. *Induction, Acceptance, and Rational Belief* ed. Marshall Swain Dordrecht: Reidel.
86. Harper, C.W.
    1980 'Relative Age Inference in Paleontology'. *Lethaia* 13, 239-248.
86a. Harrah, David
    1963 *Communication: A Logical Model*. Cambridge, Mass.: M.I.T. Press.
87. Hart, H.L.A. and A.M. Honore
    1969 *Causation in the Law*. London: Oxford University Press.
88. Heise, David R.
    1969 'Problems in Path Analysis and Causal Inferences'. *Sociological Methodology* Chapter 2. San Fransisco: Josey-Bass.

89. Helmer, Olaf
   1966 *Social Technology.* New York and London: Basic Books, Inc.
90. Hilpinen, Risto
   1970 'Can and Modal Logic'. *Ajatus* 32, 7-17.
91. Hintikka, Jaakko
   1966 'An Analysis of Analyticity'. *Deskription, Analytizitat und Existenz.* Verlag Anton Pustet, Salzburg und Munchen, 193-214.
92. Hintikka, Jaakko
   1965 'Are Logical Truths Analytic?'. *Philosophical Review,* 178-302.
93. Hintikka, Jaakko
   1966 'Are Logical Truths Tautologies?'. *Deskription, Analytizitat und Existenz* Verlag Anton Pustet, Salzburg und Munchen, 215-233.
94. Hintikka, Jaakko
   1968 'Are Mathematical Truths Synthetic *A Priori*?'. *The Journal of Philosophy* 65, 640-651.
95. Hintikka, Jaakko
   1967 '"*Cogito, Ergo Sum*: Inference or Performance?"' *Descartes: A Collection of Critical Essays* ed. by W. Doney (New York: 1967), 108-39. This article originally appeared in the *Philosophical Review* LXXI, 1962, 3-32. A revised version appears in *Meta-Meditations: Studies in Descartes (Belmont, California: 1965)*, 50-76.
96. Hintikka, Jaakko
   1981 'Information-Seeking Dialogues', in W. Becker and W.K. Essler, eds., *Konzepte der Dialektik.* Frankfurt am Main: Klostermann, 212-231.
97. Hintikka, Jaakko
   1970 'Knowledge, Belief and Logical Consequence'. *Ajatus* 32, 32-47.
98. Hintikka, Jaakko
   1962 *Knowledge and Belief.* Ithaca: Cornell University Press.
99. Hintikka, Jaakko
   1966 'Knowing Oneself' and Other Problems in Epistemic Logic. *Theoria* 32, 1-13.
100. Hintikka, Jaakko
   1970 'Surface Information and Depth Information'. *Information and Inference* ed. by Jaakko Hintikka and Patrick Suppes, Dordrecht, Reidel, 263-297.
101. Hintikka, Jaakko
   1980 'The Logic of Information-Seeking Dialogues: A Model'. *Erkenntnis* 14.

102. Hintikka, Jaakko
    1976 'The Semantics of Questions and the Questions of Semantics'. *Acta Philosophica Fennica* 28, Amsterdam, North-Holland.
102a. Hoffman, Robert
    1972 'On Begging the Question at Any Time', *Analysis* 32, 197-199.
103. Holtman, Paul D.
    1970 *The Psychology of Speakers' Audiences*. Glenview, Illinois: Scott Foresman.
104. Huff, Darrell
    1954 *How to Lie with Statistics*. New York: Norton.
105. Hughes, G.E., and M.J. Cresswell
    1968 *An Introduction to Modal Logic*. London: Methuen.
106. Hull, David L.
    1967 'Certainty and Circularity in Evolutionary Biology'. *Evolution* 21, 174-189.
106a. Hume, David
    1973 *A Treatise of Human Nature*. trans. by L.A. Selby-Bigge. Oxford: Clarendon Press.
107. Johnson Wu, Kathleen
    1973 'A New Approach for Formalization of a Logic of Knowledge and Belief'. *Logique et Analyse* 16, 513-525.
108. Johnson Wu, Kathleen
    1970 'Hintikka and Defensibility'. *Ajatus* 32, 25-31.
109. Johnson, R.H. and J.A. Blair
    1977 *Logical Self-Defense*. Toronto: McGraw-Hill Ryerson.
109a. Johnson, R.H. and J.A. Blair
    1985 'Informal Logic: The Past Five Years, 1978-1983', *American Philosophical Quarterly*, 22.
110. Johnstone, Jr., Henry W.
    1970 'Philosophy and *Argumentum ad Hominem* Revisited'. *Revue Internationale de Philosophie* 24, 107-16.
111. Kahane, H.
    1982 *Logic and Philosophy*. 4th ed., Belmont: Wadsworth.
112. Karlins, M., and H. Abelson
    1970 *Persuasion: How Opinions and Attitudes are Changed*. New York: Springer.
113. Kenny, Anthony
    1963 *Action, Emotion and Will*. London: Routledge and Kegan Paul.
114. Kneale, W. and M. Kneale
    1962 *The Development of Logic*. Oxford: Clarendon Press.

115. Krasnican, M.
    1952 'The Need for Science Classroom Procedures in Thinking'. *Science Education* 36, 123-125.
116. Kretzmann, N., (ed.)
    1966 *William of Sherwood's Introduction to Logic*. Minneapolis: University of Minnesota Press. (13th century).
117. Kripke, Saul
    1965 'Semantical Analysis of Intuitionistic Logic I'. *Formal Systems and Recursive Functions* ed., J.N. Crossley and M.A.E. Dummett. Amsterdam: North-Holland, 92-130.
118. Kruger, A.
    1964 *A Classified Bibliography of Argumentation and Debate*. New York and London: Scarecrow Press.
119. Kyburg, Henry
    1964 'Recent Work in Inductive Logic'. *American Philosophical Quarterly* 1, 250-286.
120. Larabee, Harold L.
    1954 *Reliable Knowledge*. Boston.
121. Lehrer, Keith
    1973 'Relevant Deduction and Minimally Inconsistent Sets'. *Philosophia* 3, 153-165.
122. Lehrer, Keith and Richard Taylor
    1965 'Time and Modalities'. *Mind* 74, 390-398.
123. Lesniewski, S.
    'On the Foundations of Mathematics'. *Przeglad Filozoficzny*
    1927 30, 164-206.
    1928 31, 261-91.
    1929 32, 60-101.
    1930 33, 77-105.
    1931 34, 142-170.
124. Lewis, David
    1973 'Causation'. *The Journal of Philosophy* LXX, 556-567.
125. Locke, John
    1961 *An Essay Concerning Human Understanding* ed. John Yolton, London, Dent.
126. Lorenzen, Paul
    1961 'Ein Dialogisches Konstructivtaskriterium'. *Infinitistic Methods*. London: Pergamon Press, 193-200.
127. Mackenzie, J.D.
    1979 'Question-Begging in Non-Cumulative Systems'. *Journal of Philosophical Logic*, 8, 117-133.
128. Mackie, J.L.
    1965 'Causes and Conditions'. *American Philosophical Quarterly*, vol. 2, 245-264.

129. Mackie, J.L.
     1974   *The Cement of the Universe*. Oxford: The Clarendon Press.
130. Mackie, J.L.
     1966   'The Direction of Causation'. *Philosophical Review* LXXV, 441-466.
131. Mackie, J.L.
     1964   'Self-Refutation: A Formal Analysis'. *The Philosophical Quarterly* 14, 193-203.
132. Madden, Edward H.
     1971   'Hume and the Fiery Furnace'. *Philosophy of Science* XXXVIII, 64-78.
133. Massey, Gerald
     1981   'The Fallacy behind Fallacies', *Midwest Studies in Philosophy* 6, 489-500.
134. Mavrodes, George
     1970   *Belief in God*. Random House, New York.
135. McGuire, William J.
     1960   'Cognitive Consistency and Attitude Change'. *Journal of Abnormal and Social Psychology* 60, 345-353.
136. McKinsey, J.C.C. and Alfred Tarski
     1948   'Some Theorems about the Sentential Calculi of Lewis and Heyting'. *Journal of Symbolic Logic* 13, 1-15.
137. Michalos, Alex
     1969   *Principles of Logic*. Englewood Cliffs: Prentice-Hall.
138. Mill, J.S.
     1970   *A System of Logic*. London: Longmans. Also Longmans, Green 1843.
139. Moore, G.E.
     1939   'Proof of an External World'. *Proceedings of the British Academy* v. XXV, London, Oxford University Press, 273-300. Reprinted in Robert Ammerman (ed.), *Classics of Analytic Philosophy* (1965). New York: McGraw-Hill, 68-84.
140. Moroney, M.J.
     1956   *Facts from Figures* 3rd. ed., Harmondsworth, England: Penguin Books.
141. Nagel, Ernest
     1968   'The Quest for Uncertainty'. ed., Paul Kurtz, *Sidney Hook and the Contemporary World*. New York: John Day, 407-426.
142. O'Rourke, J.E.
     1976   'Pragmatism Versus Materialism in Stratigraphy'. *American Journal of Science* 276, 47-55.

143. Perelman, Chaim and L. Olbrechts-Tyteca
    1969 *The New Rhetoric: A Treatise on Argumentation* translated by John Wilkinson and Purcell Weaver. Notre Dame and London: University of Notre Dame Press.
144. Peter of Spain
    1972 *Tractatus (Summulae Logicales)*, ed. by L.M. de Rijk, Assen, The Netherlands: Van Gorcum.
145. Penelhum, Terence
    1971 *Problems of Religious Knowledge*. Herder and Herder, New York.
146. Popper, Karl
    1968 *Conjectures and Refutations: The Growth of Scientific Knowledge*. New York: Harper and Row Torchbooks.
147. Popper, Karl
    1965 *The Logic of Scientific Discovery* 3rd ed. New York: Harper and Row Torchbooks.
148. Popper, Karl
    1965 *The Open Society and Its Enemies* II. New York: Harper and Row Torchbooks.
149. Pörn, Ingmar
    1974 'Some Basic Concepts of Action'. *Logical Theory and Semantics* ed. by Soren Stenlund, Dordrecht: Reidel, 93-101.
150. Pörn, Ingmar
    1971 *The Logic of Power*. Oxford: Blackwell.
151. Prior, A.N.
    1957 *Time and Modality*. Oxford: Oxford University Press.
152. Prior, Arthur N. and Mary Prior
    1955 'Erotetic Logic'. *Philosophical Review* 64, 43-59.
153. Quine, W.V. and J. Ullian
    1970 *The Web of Belief*. New York; Random House.
154. Quine, W.V.
    1951 *Mathematical Logic*. New York: Harper and Row, rev. ed.
155. Quine, W.V.
    1972 *Methods of Logic* 3rd. ed., New York: Holt Rinehart.
156. Rastall, R.H.
    1956 'Geology'. *Encyclopedia Brittanica* 10, 168-173.
157. Reid, Thomas
    1969 *Essays on the Intellectual Powers of Man*. Cambridge, Massachusets: M.I.T. Press. Originally published 1813-15.
158. Rescher, Nicholas
    1977 *Dialectics*. Albany: State University of New York Press.

159. Rescher, Nicholas
    1964 *Introduction to Logic.* New York: St. Martin's Press.
160. Rescher, Nicholas
    1977 *Methodological Pragmatism.* Oxford: Basil Blackwell.
161. Rescher, Nicholas
    1976 *Plausible Reasoning.* Assen/Amsterdam, Van Gorcum.
162. Rescher, Nicholas
    1973 *The Coherence Theory of Truth.* Oxford: Oxford University Press.
163. Rescher, Nicholas and Olaf Helmer
    1959 'On the Epistemology of the Inexact Sciences'. *Management Science* XI, 25-52.
163a. Routley, Richard
    1978 'The Implication Connection, and the Ensuing Inadequacy of Irrelevant Logics Such as Classical and Modal Logics', unpublished typescript.
164. Rowe, W.L.
    1962 'The Fallacy of Composition'. *Mind* 71, 87-92.
165. Russell, Bertrand
    1961 Introduction to the *Tractacus Logico-Philosophicus.* Ludwig Wittgenstein, Humanities Press, New York. p xxi-xxii.
166. Russell, Bertrand
    1913 'On The Notion of Cause'. *Proceedings of the Aristotelian Society* XIII, 1-26. Reprinted in *Mysticism and Logic* (London: Longmans Green, 1918), 180-208.
167. Salmon, Wesley
    1963 *Logic.* Englewood Cliffs, N.J., Prentice-Hall.
168. Sanford, David
    1972 'Begging the Question'. *Analysis* 32, 197-99.
169. Sanford, David
    1977 'The Fallacy of Begging the Question: A Reply to Barker'. *Dialogue* 16, 485-98.
170. Scriven, Michael
    1976 *Reasoning.* New York: McGraw-Hill.
171. Siegel, Harvey
    1985 'Educating Reason: Critical Thinking, Informal Logic and the Philosophy of Education'. *Informal Logic*, 7, 69-82.
172. Shoesmith, D. and T.J. Smiley
    1980 *Multiple-Conclusion Logic.* Cambridge-New York, Cambridge University Press.
173. Sidgwick, Alfred
    1884 *Fallacies: A View of Logic from the Practical Side.* D. Appleton & Co., New York.

174. Simon, H.A.
    1954 'Spurious Correlation: A Causal Interpretation'. *Journal of the American Statistical Association* XLIX, 467-492.
175. Simon, Herbert
    1954 'Bandwagon and Underdog Effects and the Possibility of Election Predictions'. *Public Opinion Quarterly* 18, 245-253.
176. Singer, Peter
    1972 'Moral Experts'. *Analysis* 32, 115-117.
177. Skyrms, Brian
    1966 *Choice and Chance: An Introduction to Inductive Logic*. Belmont: California: Dickenson.
178. Stalnaker, Robert C.
    1968 'A Theory of Conditionals'. *Studies in Logical Theory* ed. Nicholas Rescher. Oxford: Blackwell.
179. Stegmuller, W.
    1964 'Remarks on the Completeness of Logical Systems Relative to the Validity Concepts of P. Lorenzen and K. Lorenz'. *Notre Dame Journal of Formal Logic* 5, 81-112.
180. Suppes, P.
    1970 *A Probabilistic Theory of Causality*. Amsterdam: North-Holland.
181. Suppes, P.
    1973 'Problems in the Philosophy of Space and Time'. *Space, Time and Geometry* edited by P. Suppes. Dordrecht: Reidel Publishing Company.78
182. Tarski, Alfred
    1956 *Logic, Semantics, Metamethematics*. Oxford: Oxford University Press.
183. Tarski, Alfred
    1969 'Truth and Proof'. *Scientific American* 220, 63-77.
184. Thompson, William Irwin
    1971 *At the Edge of History*. New York: Harper and Row.
185. Turoff, Murray
    1970 'The Design of a Policy Delphi'. *Technical Memorandum* TM-123. National Resources Analysis Center, Systems Evaluation Division.
186. Van Dun, Frank
    1972 'On the Modes of Opposition in the Formal Dialogues of P. Lorenzen'. *Logique et Analyse* 57-58, 103-136.
187. Van Fraasen, Bas C.
    1970 *An Introduction to the Philosophy of Time and Space*. New York: Random House.

188. Vignaux, Georges
    1976 *L'Argumentation*. Geneve: Librairie Droz.
189. Von Wright, G.H.
    1971 *Explanation and Understanding*. Ithaca, New York: Cornell University Press.
190. Von Wright, G.H.
    1957 *Logical Studies*. Routledge and Kegan Paul, London.
191. Walton, Douglas
    1985 'Are Circular Arguments Necessarily Vicious?'. *American Philosophical Quarterly* 22, 263-274.
191a. Walton, Douglas
    1985 *Arguer's Position*. Westport, Connecticut: Greenwood Press.
192. Walton, Douglas
    1979 'Critical Study of Some Recent Action Theory'. *Philosophia* 719-740.
193. Walton, Douglas
    1984 *Logical Dialogue Games and Fallacies*. Lanham, Md.: University Press of America.
193a. Walton, Douglas
    1987 *Informal Fallacies*. Amsterdam, John Benjamins.
194. Walton, Douglas
    1976 'Logical Form and Agency'. *Philosophical Studies* 29, 75-89.
195. Walton, Douglas
    1975 *Mathematical Reviews* MR 12670, 50, 1975.
196. Walton, Douglas
    1975 'Mill and DeMorgan on Whether the Syllogism is a Petitio'. *Historia Mathematica* 2, 336-337 [abstract].
197. Walton, Douglas
    1977 'Mill and DeMorgan on Whether the Syllogism is a Petitio'. *International Logic Review* 8, 57-68.
197a. Walton, Douglas
    1985 'New Directions in the Logic of Dialogue', *Synthese*, 63, 259-274.
198. Walton, Douglas
    1977 'Performative and Existential Self-Verifyingness'. *Dialogue* 16, 128-138.
199. Walton, Douglas
    1980 '*Petitio Principii*' and Argument Analysis', in *Informal Logic* ed., J.A. Blair and R.H. Johnson. Cal.: Edgepress, 41-54.
200. Walton, Douglas
    1979 'Philosophical Basis of Relatedness Logic'. *Philosophical Studies* 36, 115-36.

200a. Walton, Douglas
1988 'Question-Asking Fallacies', *Questions and Questioning*. ed. Michel Meyer, Berlin: De Gruyter, 195-121.
201. Walton, Douglas
1973 'The Contemporary Relevance of Hume's Remarks on Liberty and Necessity'. *Journal of Thought* VIII, 183-188.
202. Walton, Douglas
1981 'The Fallacy of Many Questions'. *Logique et Analyse* 95-96, 291-313. Chapter 18 herein.
203. Walton, Douglas
1976 'Time and Modality in the 'Can' of Opportunity'. *Action Theory* ed. by Myles Brand and Douglas Walton. Dordrecht: Reidel.
203a. Walton, Douglas
1982 *Topical Relevance in Argumentation*. Amsterdam: John Benjamins.
204. Walton, Douglas
1980 'Why is The *Ad Populum* A Fallacy'. *Philosophy and Rhetoric* 13, 264-278. Chapter 16 herein.
204a. Walton, Douglas
1989 *Informal Logic*. Cambridge: Cambridge University Press.
205. Walton, Douglas and Lynn M. Batten
1984 'Games, Graphs and Circular Arguments'. *Logique et Analyse* 106, 133-164.
205a. Wang, Hao
1978 'On Formalization', in Copi and Gould *Contemporary Philosophical Logic*. New York: St. Martin's Press, 2-13.
206. Watts, I.
1775 *Logic, or the Right Use of Reason in the Enquiry after Truth, with a Variety of Rules to Guard Against Error, in the Affairs of Religion and Human Life, as well as in the Sciences*. London: Clark and Hett.
207. Weaver, Timothy W.
1971 'The Delphi Forecasting Method'. *Phi Delta Kappan*.
208. Whately, Richard
1840 *Elements of Logic*. New York: Sheldon & Co.
209. Whately, Richard
1963 *Elements of Rhetoric*. ed. Douglas Ehringer, Carbondale, Southern Illinois University Press, Part I, Ch. III. This is a reprint of seventh edition of 1828.
209a. Wheatley, J.M.O.
'Deliberative Questions'. *Analysis* 15, 49-60.

210. William of Sherwood
    1966 *Introduction to Logic*. Translated and edited by Norman Kretzmann. Minneapolis: University of Minnesota Press.
211. William of Sherwood
    1968 *Syncategorematic Terms*. Translated and edited by Norman Kretzmann. Minneapolis: University of Minnesota Press.
212. Woods, John
    1989 'Aquarian Implication: The Relevance of the Relevance Logic', in *Directions in Relevant Logic*, ed. Richard Sylvan, Dordrecht; Kluwer Academic Publishers (1989).
213. Woods, John
    1967 'Is There a Relation of Intensional Conjunction?'. *Mind* 78, N.S., 357-368.
214. Woods, John
    1965 'On How Not to Invalidate Disjunctive Syllogism'. *Logique et Analyse* 8, 312-320.
215. Woods, John
    1965 'On Arguing about Entailment'. *Dialogue* III, 405-421.
216. Woods, John
    1975 *Proof & Truth*. Toronto, Peter Martin Associates.
217. Woods, John
    1973 'Semantic Kinds'. *Philosophia* 2, 117-151.
218. Woods, John
    1974 *The Logic of Fiction: A Philosophical Sounding of Deviant Logics*. The Hague: Mouton.
218a. Woods, John
    1988 'Ideals of Rationality in Dialogic'. *Argumentation*. 2, 395-408.
219. Woods, John
    1965 'Was Achilles' 'Achilles' Heel', Achilles' Heel?'. *Analysis* 25, 142-146.
220. Woods, John
    1980 'What is Informal Logic?' *Informal Logic*, ed., R.H. Johnson and J.A. Blair. Pt. Reyes, Cal.: Edgepress, 57-68.
221. Woods, John and Douglas Walton
    1974 'Towards a Theory of Argument'. *Metaphilosophy*, 299-315.
222. Woods, John and Douglas Walton
    1976 *'Ad Baculum'*, *Grazer Philosophische Studien* 2, 133-140
223. Woods, John and Douglas Walton
    1977 *'Ad Hominem'*. *Philosophical Forum* 8, 1-20.

224. Woods, John and Douglas Walton
    1977 *'Ad Hominem*, Contra Gerber'. *Personalist* 58, 141-144.
224a. Woods, John and Douglas Walton
    1982 *Argument: The Logic of the Fallacies.* Toronto: McGraw-Hill.
225. Woods, John and Douglas Walton
    1974 *'Argumentum ad Verecundiam*'. *Philosophy and Rhetoric* 7, 135-153.
226. Woods, John and Douglas Walton
    1978 'Arresting Circles in Formal Dialogues'. *Journal of Philosophical Logic* 7, 73-90.
227. Woods, John and Douglas Walton
    1979 'Circular Demonstration and Von Wright-Geach Entailment'. *Notre Dame Journal of Formal Logic* vol. XX, 768-772.
228. Woods, John and Douglas Walton
    1977 'Composition and Division'. *Studia Logica* XXXVI, 381-406.
229. Woods, John and Douglas Walton
    1976 'Fallaciousness Without Invalidity?' *Philosophy and Rhetoric* 9, 52-54.
230. Woods, John and Douglas Walton
    1974 'Informal Logic and Critical Thinking'. *Education* 95, 84-86.
231. Woods, John and Douglas Walton
    1975 'Is the Syllogism a *Petitio Principii*?'. *Mill News Letter* X, 13-15.
232. Woods, John and Douglas Walton
    1979 'Laws of Thought and Epistemic Proofs'. *Idealistic Studies*, 55-65.
233. Woods, John and Douglas Walton
    1972 'On Fallacies'. *Journal of Critical Analysis* 4, 103-112.
234. Woods, John and Douglas Walton
    1975 *'Petitio Principii*'. *Synthese* 31, 107-127.
235. Woods, John and Douglas Walton
    1977 *'Petitio* and Relevant Many-Premised Arguments'. *Logique et Analyse* 77-78, 97-110.
236. Woods, John and Douglas Walton
    1977 *'Post Hoc Ergo Propter Hoc'*. *Review of Metaphysics* 30, 569-593.
237. Woods, John and Douglas Walton
    1978 'Puzzle for Analysis: Find the Fallacy'. *Informal Logic Newsletter* 1, 5-6.
238. Woods, John and Douglas Walton

1978 'The Fallacy of *Ad Ignorantiam*'. *Dialectica* 32, 87-99.
239. Woods, John and Douglas Walton
1982 'The *Petitio*: Aristotle's Five Ways'. *Canadian Journal of Philosophy* 12, 77-100.
240. Woods, John and Douglas Walton
1979 'What Type of Argument is an *Ad Verecundiam*?' *Informal Logic Newsletter* 2, 5-6.
241. Woods, John and Douglas Walton
1979 'Formal Logic and the Logic of Argument'. *Abstracts of the 6th International Congress of Logic, Methodology and Philosophy of Science* 5, 209-212.
242. Woods, John and Douglas Walton
1982 'Question-Begging and Cumulativeness in Dialectical Games'. *Noûs* 16, 585-606.
243. Woods, John and Douglas Walton
1979 'Equivocation and Practical Logic'. *Ratio* 21, 31-43.
244. Wright, Sewall
1934 'The Method of Path Coefficients'. *Annals of Mathematical Statistics* V, 161-215.
245. Zeisel, Hans
1968 *Say it with Figures* 5th ed., New York: Harper & Row.
246. Zeman, Zbynek
1973 *Nazi Propaganda.* New York: Oxford University Press.

Acknowledgements

Chapter 1 was originally published in the Journal of Critical Analysis, **5**, (1972), 103-111. Permission to republish is gratefully acknowledged.

Chapter 2 was originally published in Philosophy and Rhetoric, **7**, no. 3 (1974), 135-153. The Pennsylvania State University Press, University Park, P.A. Permission to republish is gratefully acknowledged.

Chapter 3 was originally published in Synthese, **31** (1975), 107-127. Copyright 1975 by D. Reidel Publishing Co. Permission to republish is gratefully acknowledged.

Chapter 4 was originally published in Grazer Philosophische Studien, **2** (1976), 133-140. Permission to republish is gratefully acknowledged.

Chapter 5 was originally published in The Philosophical Forum, **8**, 1 (1977), 1-19. Permission to republish is gratefully acknowledged.

Chapter 6 was originally published in Logique et Analyse, **77-78** (1977), 97-110. Permission to republish is gratefully acknowledged.

Chapter 7 was originally published in The Personalist, **58** (1977), 14-144. Permission to republish is gratefully acknowledged.

308 Acknowledgements

Chapter 8 was originally published in Studia Logicia, **36** (1977), 381-406. Permission to republish is gratefully acknowledged.

Chapter 9 was originally published in The Review of Metaphysics, 30, (1977), 569-593. Permission to republish is gratefully acknowledged.

Chapter 10 was originally published in the Journal of Philosophical Logic, 7 (1978), 73-90. Copyright © 1978 by D. Reidel Publishing Co. Permission to republish is gratefully acknowledged.

Chapter 11 was originally published in Dialectica, **32** (1978), 87-9. Permission to republish is gratefully acknowledged.

Chapter 12 was originally published in Notre Dame Journal of Formal Logic, **20** (1979), 768-772. Permission to republish is gratefully acknowledged.

Chapter 13 was originally published in Idealistic Studies, **9**, 1 (1979), 55-65. Permission to republish is gratefully acknowledged.

Chapter 14 was originally published in Informal Logic Newsletter, 2, 1 (1979), 5-6. Permission to republish is gratefully acknowledged.

Chapter 15 was originally published in Ratio, **21** (1979), 31-43. Permission to republish is gratefully acknowledged.

Chapter 16 was originally published in Philosophy and Rhetoric (1980), **13**, 264-278. Published by the Pennsylvania State University Press, University Park, PA. Permission to republish is gratefully acknowledged.

Chapter 17 was originally published by Informal Logic: the First International Symposium (1980), pp. 57-68. Permission to republish is gratefully acknowledged.

Chapter 18 was originally published in Logique et Analyse, **95-96**, (1981), 291-313. Permission to republish is gratefully acknowledged.

Chapter 19 was originally published in Noûs, **16**, (1982), 585-606. Permission to republish is gratefully acknowledged.

Index

Index of subjects

Abusive ad hominem 65
Abusive Terms 88
Acceptance 59
Advertising ethics 209
Aggregate-theory 107
Aggregates 98
Amphiboly 95
Analogies 206
Antecedent mutual cause 137
Applied logic xvii, 218
Approximate spatiotemporal coincidence 130
Arguing in a circle 29
Argument 2
   atomic 83
   demonstrative 56
   dialectical 8
   fallacious 1
   from authority 2
   from popularity 212
   molecular 83
   probabilistic 6
   prudential 49
   relevant 80
Argumentum
   ad baculum 48
   ad hominem 55
   ad ignorantiam 25
   ad misericordiam 213
   ad populum 209
   ad verecundiam 11
Assertion modality 59
Authority 12
   de facto 21
   de jure 21
   religious 22

312 *Index of subjects*

Befortification 43
Begging the question 29, 35
Behavioral aberration 210
Black and White Fallacy 247, 248
Bodies 104
Branching 157
Burden of proof 259
Carousel 34
Categorical assertion 259
Category mistake 102
Causal cellular automata 140
Causal lawfulness 125
Causal logic 121
Causal necessity 125
Cautious assertion 259
Chains 157
Challenge busting 264
Challenges 263
Chinese box 103
Circular sequence 155
Circularity 32
Circulus probandi 29
Circumstantial ad hominem 58
Classical first-order logic 210
Classical formal erotetic logic 237
Classical truth-functions 151
Coercive fallacies 245
Collective predication 98
Commitment 61, 155, 269
    commitment-store operation 146
    commitment-stores 148, 168
Common Law 27
Common part 104
Common sense 139
Compactness Metatheoram xvi
Competence 19
Complex question 233
Componentiation 107
Composition 93
Compositionally heridity 109
Conditional credence 44
Conditional interrogative 240
Conditioned interrogative 240
Confirmation 15
Conjunction 151
Conjunctive conclusions 83

Index of subjects 313

Consensus techniques 21
Context 200
Contextual shift 198
Conventional 102
Crackpots 70
Cranks 70
Credence 44
Credentials 12
Cumulative 154
Cumulativeness 153
Data-processing procedures 205
De Morgan's Defence 81
De Morgan's Thesis 79
De Morgan's Weaker Thesis 79
De Morgan's Zwischenzug 84
Decision theory 49
Decision-procedure 108
Deductive logic 3
Deductive paradigms 212
Deductive validity 2
Deductive-nomological 12
Deep deception 204
Defensibility 35
Delphi technique 21
Demonstration
   non-circular 175
Deontic logic 49
Deontic-prudential logical form 49
Dependency conception 30
Dependency-circularity 35
Dependent variables 134
Design and Monitor Team 25
Diachronicality 32
Dialectic 25
Dialectical equivalence 270
Dialectical precepts 8
Dialectical premiss 211
Dialectical system 61, 166
Dialectically incorrect 266
Dialogical games
   See also games of dialectic
Direct answer 236
Direct relatedness 270
Disguised premisses 31
Disjunction 151
Disjunctive Syllogism 32

314   *Index of subjects*

Dissociation 36
Distributive predication 98
Division 93
Divisionally heridity 109
Doxastic inattentiveness 37
Doxastic logic 144
Ecology 15
Economics 122
Effectiveness 43
Emotional appeal 209
Emotional turbulence 219
Emotive distraction 220
Empty aggregate 107
Engineering 122
Epistemic assumptions 183
Epistemic context 38
Epistemic logic 5
Epistemic semantical analysis of intuitionistic logic 185
Epistemic-doxastic erotetic logic 236
Epistemically incorrect 266
Equivalence conception 30
Equivalence-circularity 35
Equivalency Reduction Thesis 83
Equivocation 50, 195
Erotetic fallacies 214
Erotetic logic 235
Erotetic models 250
Euclidian geometry 105
Evidence 151, 156, 162
    direct 19, 26
    total 7
Evidentiary progression 158
Expert testimony 12
Expertise 11
Explanation
    stochastic 14
Exportation 271
Fallacy 1
Fallacy of Many Questions 233
Fallacy of Overrating 115
Fallacy within language 201
False cause 121
False presupposition 237
Feelings 217
Field 122
Finite dissection 104

Four-dimensional differentiable manifold 130
Functional relationships 129
Gambler's Fallacy 116
Games of dialectic 61, 145, 147, 155, 206, 253, 254, 261, 264, 269
Glamor 18
Greatest common part 104
Groundedness 155, 264
Hamblin's System 2 257
Horned man fallacy 249
Humean approach 123
Ideal rationality assumptions 44
Ignoratio elenchi 215
Immediate consequence 269
Implication 127, 151, 270
    virtual 35
Inarticulable expertise 19
Inconsistency 62, 63
    logical 59
Independent variable 134
Indirect relatedness 270
Individual 107
Inductive validity 2
Inference 7
    immediate 35
    plausible 192
    prudential 48
Informal fallacy 29, 201
Initial conditions 123
Initial thesis 61
Intensional disjunction 33
Interquartile range 24
Intervening variable 134
Intuitionistic 152
    calculus 154
    model structure 152
Invalidity 199
Join 104
KK-hypothesis 7
Knowledge 20
Kretzmann's Thesis 97
Kripke model 151
Kripke tree structure 151
Laudatory Ad Hominem 89
Law 27
Laws of nature 123
Layman 12

Least envelope 104
Legal system 210
Loaded questions 244
Local comparability of time order 130
Logic of dialogues xvi
Looping back 266
Lottery Fallacy 116
Many questions 233
Mass enthusiasm 220
Mechanics 104
Medicine 15
Meet 104
Mentalism 14
Mereology 104
Metalanguage 2
Metaphors 206
Methods of conditional guarding 239
Model 152
Modus ponens 7
Modus tollens 50
Monolectical theories 9
Moral expertise 23
Multiple questions 246
Multiple-choice questions 250
Natural languages 205
Natural properties 102
Necessity in nature approach 125
Negation 151, 186
    strong 172
    weak 172
Negative correlations 138
Neighbourhood 123
Neutral terms 88
Nexpertise 66
Nodes 154
Noll-Suppes theory 104
Non causa pro causa 132
Non-circular groundedness 156
Non-classical Connectives 184
Non-cumulativeness 154
Non-probabilistic theoretical laws of nature 127
Object language 2
Objective knowledge 20
Paronymous 201
Parts 101
Passions 217

Index of subjects 317

Petitio principii 2
Plausible reasoning 192
Political science 122
Positive correlation 129
Positum 253
Practical Logic 195
Practical syllogism 49
Pragmatic context 242
Pragmatic factors 233
Pragmatic relevance 126
Prestige 18
Presumptions 168
Presumptive validity 2
Presupposition 236
Principle of extensionality 107
Private languages 205
Privileged access 26
Probability calculus 44
Profortification 43
Proof 35, 182
    cantorian diagonal 13
    circular 188
    Indirect 187
Properties 102
Propositum 254
Provisoed assertion 260
Psychiatry 15
Psychologism 144
Public affairs 209
Public opinion 210
Puns 206
Quantification 36
Quantifiers 36
Quasi-laws 15, 122
Reductio ad absurdum 60, 187
Relatedness conditional connective 128
Relatedness equivalence 270
Relatedness logic 215
Relevance logic 215
Rescher Dialectics 259
Respondent group 25
Reversing cause and effect 133
Rhetoric 218
Risky question 238
Rule of addition 128
Safe questions 236

Sampling theory 138
Say-so 206
Scientific objectivity 12
Secundum quid 132
Self-sustaining 33
Self-sustenance 35
Semantic closure 205
Semantic consequence 33
Semantic kind properties 108
Semantic relatedness 128
Semantics 205
Set theory 104
Sheffer-stroke xvi
Simultaneous causation 125
Single questions 246
Sociology 218
Soundness 43
Speculation 20
Standard Treatment 47
Stipulated meaning 204
Strong intensionality 33
Structure of bodies 104
Subjunctive conditionals 124
Superknower 5
Suppressing information 138
Syntactic ambiguity 202
Syntax 205
Tertullian 53
Testability 169
Thaumatrope 115
Threat of force 50
Topic neutrality 69
Topic relevance 216
Tortoise 38
Transitive closure 270
Translation 197
Transumption 200
Trees 150, 154
Truth functionality 33
Truth-creation 28
Tu quoque 55
Unprojectible predicates 106
Unsoundness 199
Unwarrantedness 243
Unwelcome presupposition 241
Unwelcomeness 243

User Body 25
Validity 2, 35
Variance 134
Verified 186
Violence 51
weak intensionality 33
Weakly fortifies 43
Whately-Prior method 239
Whether-if method 240
Whether-question 236
Whether-whether method 238
Wholes 101
World alternativeness 124
World-similarity 124
Yes-no questions 236

Index of persons

Achilles 38
Anderson, A.R. 61
Anderson, John 122
Aristotle 29
Arnauld, Antoine 57
Bar-Hillel, Y. 111
Barker, John A. 144
Barth, E.M. xx
Belnap, Nuel D. Jr. 61
Bentham, Jeremy 42
Berger, George 131
Binkley, Robert xviii
Blair, Anthony xx
Bochenski, I.B. 96
Boole, George 77
Bosanquet, Alfred 181
Bradley, F.H. 181
Burge, Tylor 108
Burks, Arthur 125
Burleigh 96
Carlson, Laurie xx
Carnap, Rudolf 111
Chisholm, Rodrick 124
Chomsky, Noam 14
Cole, K.R. 112
Copi, I.M. 53
Cresswell, M.J. 117
Dalkey, N. 20
Ditmarsch, Hans van xxi
Domotor, Zolton 127
Ducasse, C.T. 123
Dun, Frank van 148
Eemeren, Frans van xx
Ellul, Jacques 217
Epstein, R.L. 128
Fischer, David H. 137, 248
Finocchiaro, Maurice A. 56
Frege, Gottlob 77

Index of persons 321

Geach, Peter 175
Gerber, D. 87
Goldblatt, R.I. 128
Goodman, Nelson 124
Govier, Trudy xx
Green, Romuald 253
Grootendorst, Rob xx
Hamblin, C.L. 1
Hanfstaengl, Ernst 52
Harman, Gilbert 7
Harrah, David 238
Helmer, Olaf 14
Hempel, C.G. 12
Hintikka, Jaakko xx, 39
Hitchcock, David xx
Hitler, Adolf 52
Hughes, H.E. 117
Hume, David 123
John Deegan, Jr. 134
Johnson Wu, Kathleen 143
Johnson, Ralph xx
Johnstone, Henry W. Jr. 57
Kneale, M. 94
Kneale, W. 94
Krabbe, Erik xx
Kretzmann, Norman 96
Kripke, Saul 61
Lehrer, Keith 61
Lennart Åqvist 233
Lesniewski, S. 104
Lewis, David 124
Locke, John 55
Loewen, David 139
Lorenz, Kuno xx
Lorenzen, Paul 143
MacKenzie, Jim xx
Mackie, J.L. 62
Madden, Edward H. 123
Mann, William C. xxi
McKinsey, I.C.C. 185
Meyer, Michel xx
Michalos, Alex 49
Mill, J.S. 2
Moore, G.E. 41
Morgan, Augustus De 17
Peter of Spain 96

Plato 236
Prior, A.N. 238
Prior, Mary 238
Quine, W.V. 198
Reichenbach, Hans 130
Reid, Thomas 123
Rescher, Nicholas 14
Rowe, William L. 111
Russell, Bertrand 128
Salmon, Wesley 18
Sanford, David 144
Sherwood, William of 96
Sidgwick, Alfred 42
Sobel, Jordan Howard 124
Socrates 23, 236
Stalnaker, Robert 124
Stegmüller, W. 148
Strawson, P.F. 241
Suppes, Patrick 104
Swartz, Norman 131
Tarski, Alfred 181, 185
Thomas B. Steel, Jr. 234
Von Wright, Georg 126
Walton, Douglas 128
Whately, Richard 29
William of Sherwood 96
Woods, John 124
Wright, Georg H. von 175
Zeisel, Hans 136

www.ingramcontent.com/pod-product-compliance
Ingram Content Group UK Ltd.
Pitfield, Milton Keynes, MK11 3LW, UK
UKHW021317180426
11947UKWH00015B/1274